Optics

The science of light

Optics

The science of light

Paul Ewart

Clarendon Laboratory, Oxford University, Oxford, United Kingdom

Morgan & Claypool Publishers

Rights & Permissions
To obtain permission to re-use copyrighted material from Morgan & Claypool Publishers, please contact info@morganclaypool.com.

ISBN   978-1-64327-676-2 (ebook)
ISBN   978-1-64327-673-1 (print)
ISBN   978-1-64327-674-8 (mobi)

DOI   10.1088/2053-2571/ab2231

Version: 20191001

IOP Concise Physics
ISSN 2053-2571 (online)
ISSN 2054-7307 (print)

A Morgan & Claypool publication as part of IOP Concise Physics
Published by Morgan & Claypool Publishers, 1210 Fifth Avenue, Suite 250, San Rafael, CA, 94901, USA

IOP Publishing, Temple Circus, Temple Way, Bristol BS1 6HG, UK

To Katharine and Adam

Contents

Preface   x

Acknowledgements   xi

Author biography   xii

1   Introduction and structure of the course   1-1

2   Geometrical optics   2-1
2.1   Fermat's principle   2-1
2.2   Lenses and principal planes   2-2
2.3   Compound lens systems   2-3
      2.3.1 Telephoto lens   2-3
      2.3.2 Wide-angle lens   2-4
      2.3.3 Telescope (astronomical)   2-4
      2.3.4 Telescope (Galilean)   2-5
      2.3.5 Telescope (Newtonian)   2-6
      2.3.6 Compound microscope   2-7
2.4   Illumination of optical systems   2-8

3   Waves and diffraction   3-1
3.1   Mathematical description of a wave   3-1
3.2   Interference   3-2
3.3   Phasors   3-2
3.4   Diffraction from a finite slit   3-3
3.5   Diffraction from a finite slit: *phasor treatment*   3-5
3.6   Diffraction in two dimensions   3-7

4   Fraunhofer diffraction   4-1
4.1   Fraunhofer diffraction   4-1
4.2   Diffraction and wave propagation   4-3

5   Fourier methods in optics   5-1
5.1   The Fresnel–Kirchhoff integral as a Fourier transform   5-1
5.2   The convolution theorem   5-2
5.3   Some useful Fourier transforms and convolutions   5-3
5.4   Fourier analysis   5-5

5.5	Spatial frequencies	5-6
5.6	Abbé theory of imaging	5-7
5.7	Spatial resolution of the compound microscope	5-9
5.8	Diffraction effects on image brightness	5-10

6	**Optical instruments and fringe localisation**	**6-1**
6.1	Division of wavefront	6-1
	6.1.1 Two-slit interference, Young's slits	6-1
	6.1.2 N-slit diffraction, the diffraction grating	6-1
6.2	Division of amplitude	6-2
	6.2.1 Point source	6-3
	6.2.2 Extended source	6-4

7	**The diffraction grating spectrograph**	**7-1**
7.1	Interference pattern from a diffraction grating	7-1
	7.1.1 Double slit, $N = 2$	7-1
	7.1.2 Triple slit, $N = 3$	7-2
	7.1.3 Multiple slit, $N = 4$, etc	7-3
7.2	Effect of finite slit width	7-4
7.3	Diffraction grating performance	7-5
	7.3.1 The diffraction grating equation	7-5
	7.3.2 Angular dispersion	7-5
	7.3.3 Resolving power	7-6
	7.3.4 Free spectral range	7-8
7.4	Blazed (reflection) gratings	7-8
7.5	Effect of slit width on resolution and illumination	7-9

8	**The Michelson (Fourier transform) interferometer**	**8-1**
8.1	Michelson interferometer	8-1
8.2	Resolving power of the Michelson spectrometer	8-3
8.3	The Fourier transform spectrometer	8-5
8.4	The Wiener–Khinchin theorem	8-6
8.5	Fringe visibility	8-7
	8.5.1 Fringe visibility and relative intensities	8-7
	8.5.2 Fringe visibility, coherence and correlation	8-7

9	**The Fabry–Pérot interferometer**	**9-1**
9.1	The Fabry–Pérot interference pattern	9-2

9.2	Observing Fabry–Pérot fringes	9-3
9.3	Finesse	9-4
9.4	The instrument width	9-5
9.5	Free spectral range, FSR	9-5
9.6	Resolving power	9-6
9.7	Practical matters	9-7
	9.7.1 Designing a Fabry–Pérot	9-8
	9.7.2 Centre spot scanning	9-8
	9.7.3 Limitations on finesse	9-9
9.8	Instrument function and instrument width	9-9

10 Reflection at dielectric surfaces and boundaries — **10-1**

10.1	Electromagnetic waves at dielectric boundaries	10-1
10.2	Reflection properties of a single dielectric layer	10-3
10.3	Anti-reflection coatings	10-5
10.4	Multiple dielectric layers: matrix method	10-6
10.5	High reflectance mirrors	10-7
10.6	Interference filters	10-7
10.7	Reflection and transmission at oblique incidence	10-8
	10.7.1 Reflection and transmission of p-polarized light	10-8
	10.7.2 Reflection and transmission of s-polarized light	10-9
10.8	Deductions from Fresnel's equations	10-10
	10.8.1 Brewster's angle	10-10
	10.8.2 Phase changes on reflection	10-11
	10.8.3 Total (internal) reflection and evanescent waves	10-12

11 Polarized light — **11-1**

11.1	Polarization states	11-1
	11.1.1 Case 1: linearly polarized light, $\delta = 0$	11-2
	11.1.2 Case 2: circularly polarized light, $\delta = \pm\pi/2$	11-2
	11.1.3 Case 3: elliptically polarized light	11-4
11.2	Transformation and analysis of states of polarization	11-7
11.3	Optics of anisotropic media; birefringence	11-8
11.4	Production and manipulation of polarized light	11-12
	11.4.1 Modifying the polarization of a wave	11-12
	11.4.2 Production of polarized light	11-14
11.5	Analysis of polarized light	11-15
11.6	Interference of polarized light	11-17

Preface

The science of light has a long history. According to the book of Genesis, light was the first thing that God made after the initial formless, chaotic, and *dark* void. 'And God saw that it was good'. We also find it good and the study of light remains to this day an important part of human creative activity. Optical devices have been developed to image very distant astronomical objects and very small, microscopic organisms here on Earth. These have enabled our growing understanding of the Universe and the mechanisms of life itself. In particular, the analysis of the spectrum of light emitted and absorbed by matter underpinned the development of modern atomic theory and quantum mechanics, which has led to our present understanding of the fundamental quantum nature of light. This book, however, is based completely on the classical electromagnetic theory of light as a wave.

The content and structure of the book has been determined by their genesis as lecture notes accompanying a course on optics that I gave to second year undergraduates at the University of Oxford—a component of the Final Year examination syllabus. As such, it assumes a basic familiarity with the mathematics of differential equations, integration and complex numbers.

Given that the book will treat light as a wave, the basic approach is to examine the characteristic property of waves which is interference. However, a brief summary of geometrical optics, which ignores the wave nature of light, is given at the beginning, in order to help understand the operation of lenses, mirrors, etc, that will be encountered later. The mathematical treatment of waves is introduced using scalar diffraction theory, i.e. ignoring the vector nature of the electromagnetic field, and both analytical and phasor methods are described for the solving of diffraction problems in some simple cases. This theory is shown to provide a means of describing wave propagation using the Fresnel-Kirchoff diffraction integral which leads to Fourier methods for treating diffraction in more complex situations. The following parts of the book describe the physics underlying some important instruments for spectral analysis; the diffraction grating spectrograph, the Michelson and Fabry-Pérot interferometers. These optical devices, and many others, involve reflections of light at surfaces and so the physics of reflection and transmission phenomena at boundaries of dielectric media is treated next. Finally, the discussion returns to the vector nature of the wave to describe polarization effects, i.e. the effect of the direction of the electric and magnetic fields has on the propagation and reflection of light.

It is worth noting that the book is part of a series of concise texts and so, in the interests of being as concise as is consistent with clarity, the discussion has been limited to presenting the basic principles of optics and some important spectroscopic instruments. A number of other currently important topics have been omitted such as lens design, lasers, adaptive optics, etc. I hope, however, that the book will be found useful in explaining the underlying principles of optics and of the operation of a few of the more important instruments, such as to provide a solid foundation for understanding more advanced topics and applications.

Acknowledgements

This book is based on my lecture notes on Optics for the Final Honours Examination in Physics at Oxford University. I am grateful to the students who asked questions during and after my lectures. These questions and ensuing post-lecture discussions alerted me to commonly encountered difficulties that I tried to address in later versions of the notes and in this book. I have also benefited from comments from colleagues in the Physics Department arising from their own interactions with students in tutorials based on the lecture course. In particular, it is a pleasure to acknowledge the contribution of Professor Derek Stacey and Dr Geoffrey Brooker to my understanding of the subject and to Professor Armin Reichold for his helpful questions and comments on an earlier version of the notes. To both students and colleagues, I owe thanks for helping me to remove various obscurities, or even mistakes, and for making the notes—and this book—better than they would otherwise have been. Of course, I take full responsibility for any remaining deficiencies in the text.

I am grateful also to Nicki Dennis and Chris Benson, at IOP Publishing, for the invitation to contribute to this series of concise texts and for their patience and encouragement during the production of the book.

Author biography

Paul Ewart

Paul Ewart obtained a BSc and PhD in Physics from Queen's University Belfast and then was an (SERC) Advanced Fellow at the Blackett Laboratory, Imperial College of Science and Technology in London. In 1979 he moved to the Physics Department, Clarendon Laboratory, Oxford University as a lecturer and as a Tutor and Fellow of Worcester College. He has been a Royal Academy of Engineering Senior Research Fellow, a Visiting Fellow at the Joint Institute of Laboratory Astrophysics and Visiting Professor at the University of Colorado in Boulder, USA, a CNRS visiting Fellow at the Ecole Normale Superieure, Paris and a William Evans Visiting Fellow at the University of Otago, New Zealand. His research work has centred on using lasers to study atomic and molecular physics, quantum optics and nonlinear spectroscopy. His current research includes interdisciplinary applications of laser spectroscopy to combustion and environmental physics. He was Professor of Physics and formerly Head of the Department of Atomic and Laser Physics at Oxford University.

IOP Concise Physics

Optics

The science of light

Paul Ewart

Chapter 1

Introduction and structure of the course

The study of light has been an important part of science from its beginning. From early times optics has been instrumental for the technology of measurement. Eratosthenes (276–194 BC) famously measured the circumference of the Earth using the length of shadows cast by a vertical rod at two distant points on a north–south line in what is modern-day Egypt[1]. He correctly assumed that the Sun was sufficiently distant that its rays would be parallel at the Earth's surface and that the light travelled in straight lines. This principle, that light travels in straight lines, was first articulated by Heron of Alexandria (c. 10–75 AD) but has come to be known as Fermat's Principle and underlies what we call geometrical optics. Later, Islamic scholars made significant contributions with Alhazen (Ibn al-Haytham, b. 965 AD), in particular, demonstrating this rectilinear propagation of light and performing experiments using lenses and mirrors to study refraction and reflection. Medieval philosophers built on the writings of both Greek and Arabic scholars with Roger Bacon (1219–1292) working in Oxford and Paris making quantitative studies of refraction at spherical surfaces, following the path of light through lenses and, in some sense, foreseeing the operation of the telescope. With the coming of the Scientific Revolution in the 16th and 17th centuries, optics, in the shape of telescopes and microscopes, provided the means to study the Universe from the very distant to the very small. Isaac Newton (1642–1726) introduced a scientific study also of the nature of light itself, considering it to be composed of a stream of tiny 'corpuscles' or particles rather than some form of wave as suggested previously by Christiaan Huygens (1629–1695). Subsequently, this particle-picture was replaced by the wave theory supported, in particular, by the experimental observations by Thomas Young (1773–1829) and mathematical formulations by Augustin-Jean Fresnel (1788–1827)

---

[1] He obtained a value of 250 000 stadia but, sadly, since we are not certain how his unit of length, stadia, corresponds to one metre, we can't be sure how accurate his measurement was, but it was probably correct to within about 20%.

doi:10.1088/2053-2571/ab2231ch1

and others. The triumph of wave theory was seemingly sealed by James Clerk Maxwell (1831–1879) whose theory unified electricity and magnetism and indicated light to be composed of waves in the electromagnetic field. The dominance of the wave theory of light was short-lived. In 1921 Albert Einstein (1879–1955) was awarded the Nobel Prize in Physics partly for his explanation of the photo-electric effect in terms of essentially a 'particle-picture,' whereby light was composed of discrete quanta, or photons, containing a definite amount of energy. This quantum idea was consistent with the law discovered earlier by Max Planck (1858–1947) relating the energy of radiation to its frequency in discrete units. Planck's discovery explained the distribution of spectral energy radiated by 'black bodies', especially in the ultra-violet region, where classical electromagnetic theory had catastrophically failed.

Today, optics remains a key element of modern science, not only as an enabling technology, but in quantum optics, as a means of testing our fundamental understanding of quantum theory and the nature of reality itself. Since the establishment of the quantum theory of light, it is often stated that light behaves sometimes as a wave and sometimes as a particle. Arguably, a more correct view is that light is neither a wave nor a particle but better considered as a quantum entity that displays 'wave-like' or 'particle-like' properties depending on the method of detection. The 'real' nature of light remains, to some extent, an intriguing mystery!

In what follows we will be concerned principally with the wave nature of light. Most experimental studies of light will involve optical elements such as lenses and mirrors and so an understanding of how they work will be useful. We will therefore begin with a very brief summary of some relevant geometrical optics in chapter 2, which ignores the wave nature of light. We will, however, return later to see how the wave nature affects the performance of some important optical instruments.

The main focus of the book is physical optics where the primary characteristic of waves viz *interference*, is the dominant theme. It is interference that causes diffraction—the bending of light around obstacles. Our study of how waves behave will rely on having a clear mathematical description of wave phenomena and so this is introduced in chapter 3 together with a brief résumé of elementary diffraction effects. The basics of scalar diffraction theory and Fraunhofer diffraction are then introduced in chapter 4. By using a scalar theory we ignore the vector nature of the electric field of the wave, but we return to this aspect at the end of the course when we think about the polarization of light. Scalar diffraction theory allows us to mathematically treat the propagation of light and the effects of obstructions or restrictive apertures in its path. We then introduce, in chapter 5, a very powerful mathematical tool, the Fourier transform, and show how this can help in solving difficult diffraction problems. Fourier methods are used very widely in physics and recognise the inter-relation of variables in different dimensions such as 'time and frequency' or 'space and spatial frequency'. The latter concept will be useful to us in understanding the formation of images in optical systems.

Having established the mathematical basis for describing light we turn to methods of analysing the spectral content of light. The spectrum of light is the primary link between optics and atomic physics and other sciences such as astrophysics. The basis for almost all instruments for spectral analysis is, again, interference and so in chapter 6 we look at the physics underlying the formation of the interference patterns in various types of instrument. The familiar Young's slit, two-beam, interference effect in which the interference arises from division of the wavefront is generalised to multiple slits in the diffraction grating spectrometer and this is treated in chapter 7. The alternative method of producing interference, by division of amplitude, is then considered in chapter 8 where, again, we begin with the case of two beams: the Michelson interferometer and move on, in chapter 9, to multiple-beam interference in the Fabry–Pérot interferometer. These devices are important tools and play a key role in modern laser physics and quantum optics. The reflection and transmission of light at boundaries between dielectric media is an important feature of almost all optical instruments and so we then consider, in chapter 10, how the physics of wave reflection at boundaries can be engineered to produce surfaces with high or partial reflectivity or even no reflectivity at all. Finally, in chapter 11, we return to the vector nature of the electric field in the light wave. The direction in which the E-field points defines the polarization and we will study how to produce, manipulate and analyse the state of polarization of light.

IOP Concise Physics

Optics
The science of light
Paul Ewart

Chapter 2

Geometrical optics

2.1 Fermat's principle

As we noted above, the Greek thinker, Heron of Alexandria, was the first to articulate what has come to be known as Fermat's principle. Fermat, stated his principle as 'Light travelling between two points follows a path taking the least time.' The modern, and more strictly correct version, is as follows:

> *'Light propagating between two points follows a path, or paths, for which the time taken is an extremum.'*

The principle has a theoretical basis in the quantum theory of light that avoids the question of how the light 'knows' what direction to go in so that it will follow the maximal path! (Basically the wave function for the light consists of all possible paths but all, except the one corresponding to the classical path, destructively interfere owing to variations in the phase over the different paths.)

Fermat's principle may be used to derive Snell's laws of reflection and refraction. We can consider a set of possible paths for light emitted at a point, O, and travelling to a general point, P, in the region of a reflecting surface lying at vertical distance, h, from O, as shown in figure 2.1. Apart from the 'direct' straight line OP, a range of possible paths involve a reflection at a point A, a distance x, from the point on the surface directly below.

The optical path length OAP $= L$, given by,

$$L = (x^2 + h^2)^{1/2} + [(d - x)^2 + h^2]^{1/2}.$$

For a maximum or minimum,

$$\frac{dL}{dx} = 0$$

from which we find $x = d/2$.

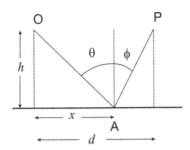

Figure 2.1. Diagrammatic representation of a general path for light propagating from a point O to point P by reflection from a point A on a surface. h is the perpendicular distance to the surface from O and P. θ and ϕ are the angles of incidence and reflection at point A.

Hence the incidence angle, θ = reflection angle, ϕ, which is Snell's law of reflection. Using a similar procedure we can derive Snell's law of refraction,

$$n_1 \sin \theta_1 = n_2 \sin \theta_2 \qquad (2.1)$$

where θ_1 and θ_2 are the angles between the light ray and the normal to the surface between media of refractive index n_1 and n_2, respectively.

Geometrical optics uses the effective rule of thumb that light travels in straight lines in a homogeneous medium of uniform refractive index. Deviations occur at boundaries between media of different refractive index or if the index varies in space. The path of light indicated by a ray can be plotted using Fermat's principle or its more useful form as Snell's laws. This allows us to locate images of objects formed when light travels through complicated lens systems or, in the case of mirages, through a medium of spatially varying refractive index.

2.2 Lenses and principal planes

It is well-known that a lens will cause any ray travelling parallel to its axis to bend and so cross the axis at a certain distance along the axis. This applies not only to a simple, single lens but to any compound lens made up of any number of lenses in sequence. Any ray entering the lens, or lens system, parallel to the axis will emerge at some angle that will determine the point at which it crosses the axis. The ray may, in a complicated system, suffer several refractions at the surfaces of each lens, each of which changes the direction of the ray. The overall effect, i.e. the refraction that has caused the resultant 'focussing', can be considered to have taken place at one point in a plane called the 'first principal plane' as shown in figure 2.2. The plane where the emerging ray crosses the axis, i.e. is focussed, is called the 'back focal plane'.

In the same way, a ray travelling through the lens system in the opposite direction will be focussed in the 'front focal plane' and this defines also the 'second principal plane' as shown in figure 2.2.

In the case where the lens can be assumed to have zero thickness we can define the focal length, f, in terms of the distance from an object to the lens, u, (object distance) and from the lens to the image, v, (image distance). The 'thin lens formula' relates the focal length to the object and image distances,

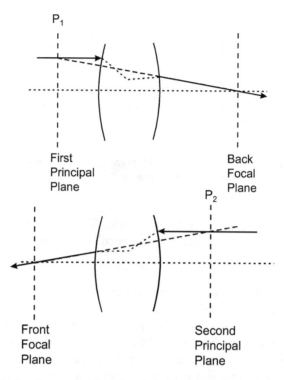

Figure 2.2. Ray diagrams defining the principal and focal planes of a lens system.

$$\frac{1}{u} + \frac{1}{v} = \frac{1}{f}. \tag{2.2}$$

u, v and f are measured along the axis to the centre of the lens. If the object is at infinity, i.e. $u = \infty$, parallel rays are focussed in the image plane where $v = f$, and this defines the focal plane of a thin lens.

For a thick or compound lens (composed of several individual lenses) the principal planes locate the position of an equivalent thin lens (see figure 2.2). The effective focal length is the distance from the principal plane to the associated focal plane.

2.3 Compound lens systems

The concepts of principal and focal planes are perhaps best illustrated by some common examples of compound lens systems.

2.3.1 Telephoto lens

Figure 2.3 shows a simple diagram of a telephoto lens, such as is used for photography, but not to scale! The combination of the positive (convex) and negative (concave) lens causes parallel rays to focus in the focal plane. For distant objects the focal plane will be very close to the image plane, i.e. where the film or

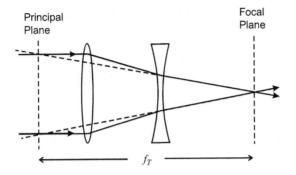

Figure 2.3. Telephoto lens showing the position of the principal plane and the focal plane separated by a distance, f_T, the effective focal length (see text).

detector in the camera will be located. The system behaves in the same way as if a single lens, with a long focal length, f_T, was placed at the principal plane as shown. The distance from the focal plane to the compound lens can be much smaller than the distance from the principal plane where the equivalent single lens would have to be located. This allows the telephoto lens to be physically much more compact than the equivalent single lens system.

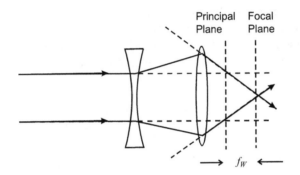

Figure 2.4. Wide-angle lens showing the position of the principal plane and the focal plane separated by a distance, f_W, the effective focal length (see text).

2.3.2 Wide-angle lens

Figure 2.4 shows a wide-angle lens where the compound lens system produces the effect of a very short focal length lens. The distance between the back surface of the lens system and the focal plane, where the image is formed, is, in this case, longer than the effective focal length, f_W. This added separation between the lens and the position of the film or photodetector in the camera provides space for the aperture and shutter system.

2.3.3 Telescope (astronomical)

The refracting astronomical telescope shown in figure 2.5 consists of an objective lens, having a focal length, f_O, and an eyepiece lens of focal length, f_E, where $f_O \gg f_E$.

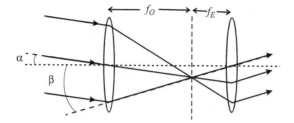

Figure 2.5. Astronomical telescope consisting of objective and eyepiece lenses with focal length given by f_O and f_E, respectively.

The objective lens forms an image of a distant object (essentially at infinity) in the back focal plane at distance, f_O, from the lens. This image acts as an 'object' for the eyepiece lens such that a virtual image of this 'object' is formed at infinity. This allows the eye to view the image in a relaxed state for ease of use.

From the geometry of the system we see that the rays from a point on the distant object enter the system at angle α and leave at an angle β. The angular magnification, M, is simply given by the ratio of these angles,

$$M = \beta/\alpha = f_O/f_E. \tag{2.3}$$

Unfortunately the image will be upside-down relative to the object! For astronomical objects this is not much of an inconvenience but it would be a bit troublesome to use the device for observing objects on Earth. This minor problem is addressed by an alternative design, attributed to Galileo, which produces an image the 'right way up'.

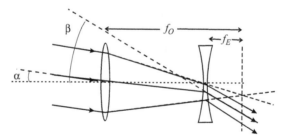

Figure 2.6. Galilean telescope consisting of objective (positive) and eyepiece (negative) lenses with focal length given by f_O and f_E, respectively.

2.3.4 Telescope (Galilean)

Figure 2.6 shows Galileo's design for a telescope using a negative (concave) lens as the eyepiece. In this arrangement the real image formed by the positive objective lens falls in the back focal plane of the eyepiece lens and so forms a 'virtual' object for this negative lens. The resulting 'virtual image' is formed by the eyepiece at infinity as shown by the dotted lines in the figure. By way of a reminder—a 'real image' is one formed in a plane where the light rays actually fall, i.e. the image would be visible if a screen were to be place in this plane. On the other hand, a 'virtual' image is in a plane where the rays do not actually fall and would not be visible on a screen in this position. By the same token, a 'virtual object' is one from which the rays *appear* to come but do not actually originate. From the geometry we see that the angular magnification, M, is again given by,

$$M = \beta/\alpha = f_O/f_E.$$

2.3.5 Telescope (Newtonian)

Both the Astronomical and the Galilean telescopes produce their magnified images as a result of refraction in the objective lens and so are known, collectively, as refracting telescopes. The degree of refraction, i.e. the bending of the light on its passage across the lens surfaces, depends on the refractive index of the glass. A problem arises because the refractive index is different for different wavelengths of light. The focal length of the lens will therefore be different for different wavelengths causing the images at different wavelengths to form at different distances from the lens. This effect is known as chromatic aberration and causes the image to be blurred with coloured edges and reduces the effective resolution obtainable by the system. This problem was addressed by Newton who used a reflecting mirror, with a spherical reflecting surface, to form the image to be viewed by the eyepiece, as shown in figure 2.7. The image formed by the spherical mirror would be formed on the axis of the mirror and it would therefore be difficult to view by eye without obstructing the incoming light. To overcome this, a small plane mirror is placed on the axis to divert the rays forming the image sideways and out of the path of the incoming light. This real image can then be viewed using the eyepiece as shown.

Once again, from the geometry of the system we find the angular magnification is,

$$M = \beta/\alpha = f_O/f_E.$$

The focal length of the objective mirror, f_O, and generally for a spherical mirror surface, equals half the radius of curvature.

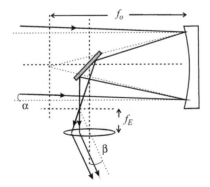

Figure 2.7. Newtonian reflecting telescope. The objective is a curved (spherical or parabolic) mirror with focal length, f_O, and the eyepiece, with focal length, f_E, is used to view the image formed by the objective reflected off a flat mirror on the mirror axis.

Chromatic aberration will still, potentially, be present in the eyepiece but this can be mitigated significantly by using an *achromatic lens*. This is usually composed of a positive and a negative lens cemented together such that the combination still forms a positive lens. When the two lenses are made of different types of glass having different dispersion properties the chromatic aberration effect can be effectively cancelled. In practice the single lenses shown in the diagrams above for the telephoto and wide-angle lenses will usually be composed of such achromatic doublet lenses. Additional lenses and specially shaped lens surfaces are also often used to correct other types of aberrations occurring in these systems.

The chief advantage of the Newton reflecting telescope over a refracting device, having the same magnification, is that it is much easier to make the objective from a large spherical mirror than a large achromatic lens. It is important, especially for astronomy, to have a large aperture system to collect more light and so make a brighter image from faint, distant objects.

2.3.6 Compound microscope

The apparent size of an object is determined by the angle it subtends at the eye, represented by the angle, α, in figures 2.5 and 2.6. To examine a very small object one would want to bring it as close as possible to the eye, but the eye has difficulty imaging anything closer than the '*near point*' of about 25 cm for normal eyesight. A compound microscope, shown in figure 2.8, forms a real image of an object of height, h, placed at a distance, u, from its objective lens having a focal length f_O. This object, when placed at the near point, a distance, D, from the eyepiece, would subtend an angle α. The object distance will be just a little bit larger that the focal length and so a magnified real image, of height h', will be formed at the image distance v, with linear magnification v/u, which can be seen from the geometry in figure 2.8.

This real image acts as an object of which the eyepiece then forms an image at infinity, which is more easily viewed by the relaxed eye. We adopt the conventional definition of the angular magnification, M, by comparing the angle subtended by the final virtual image at infinity, β, to the angle subtended by the original object if it was

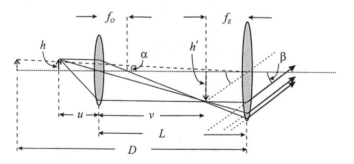

Figure 2.8. A compound microscope consists of an objective lens with a short focal length f_O forming an image at distance v, of an object at distance u. The eyepiece, focal length f_E, acts as a magnifying lens to view this real image. The separation of the objective and eyepiece is L, the length of the tube holding the lenses, and D is the distance to the 'near point' from the eyepiece (see text).

placed at the near point of the eye, α. From the geometry shown in figure 2.8, the angular magnification is,

$$M = \beta/\alpha.$$

We can express this magnification approximately in terms of the dimensions of the microscope as follows.

The linear magnification by the objective is given by,

$$\frac{h'}{h} = \frac{v}{u}.$$

We see also, from figure 2.8,

$$\alpha = \frac{h}{D} \quad \text{and} \quad \beta = \frac{h'}{f_E}$$

$$M = \frac{\beta}{\alpha} = \frac{h'}{f_E}\frac{D}{h} = \frac{v}{u}\frac{D}{f_E}.$$

Approximately, $u = f_O$ and $v = L$, so we can write,

$$M = \frac{LD}{f_O f_E}. \tag{2.4}$$

This gives a good approximation to the magnification in terms of the length of the microscope tube and the focal lengths of the lenses.

2.4 Illumination of optical systems

The brightness of an image formed by a lens will depend upon the amount of light passing through the aperture and so upon the area of the lens. A figure of merit (a dimensionless number) characterising the brightness of the image formed is given by the ratio of the focal length to the diameter, since the area will scale with the square of diameter. This figure is the 'f-number',

$$f/no. = \frac{\text{focal length}}{\text{diameter of aperture}}.$$

Aperture stops determine the amount of light reaching the image of an image forming optical system.

For lenses of given focal length, the size (diameter) of the lenses are chosen to match the aperture of the image recording system, e.g. the pupil of eye. Aperture stops for a telescope (or binocular) system will be chosen as follows:
  (a) As shown in figure 2.9, the aperture stop (eyepiece) should be approximately equal to the pupil of the eye (~10 mm).
  (b) Similarly, the aperture stop (objective) should be chosen such that the size of the image of the objective formed by the eyepiece is also approximately equal to the pupil of the eye.

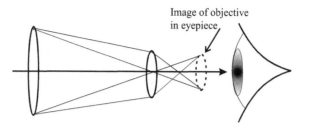

Figure 2.9. Diagram showing that the size of the objective determines the effective aperture stop in a telescope (see text).

There is no point in having an eyepiece diameter much larger than the eye's pupil since the eye, placed up close to the eyepiece, will accept light only from an area about the same size as the pupil. For the same reason there is no point in having an objective lens larger than a certain size. This limiting size is such that the eyepiece will form an image about the same size as the pupil, since the pupil can view only the small area defined by its aperture.

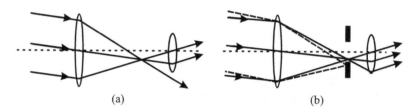

(a) (b)

Figure 2.10. (a) Some off-axis light spills past the eyepiece and is missing from the image. (b) A field stop limits the range of angles accepted but provides even illumination to the image formed by the eyepiece.

Aperture stops for a camera lens are set by an iris diaphragm over a range of *f/no.* from 1.2 to 22. A small *f/no.* is a large aperture and allows in more light. Large aperture lenses are useful for imaging in dim light or to allow a shorter shutter time to be used whilst still allowing sufficient light energy to form a recordable image. This is useful for capturing images of moving objects. The disadvantage of using large apertures is a reduced depth of field (range of distances in focus in the image).

Uneven illumination of an image results when the eyepiece does not collect all of the light incident at larger angles. Some of the off-axis light in the telescope in figure 2.10(a) spills past the eyepiece and so the light to the image at this angle is reduced in intensity, the image darkens towards the edge. In figure 2.10(b) all light at a given angle through the objective reaches the image. There is then equal brightness across the image at the expense of the field of view. A field stop, as shown in figure 2.10(b), ensures uniform illumination across the image by eliminating light at large off-axis angles.

Optics
The science of light
Paul Ewart

Chapter 3

Waves and diffraction

3.1 Mathematical description of a wave

A wave is a periodically repeating disturbance in the value of some quantity. In the case of light the oscillation is of the strength or amplitude of the electric and magnetic field. The wave amplitude, u, is shown graphically in figure 3.1 as a function of time, t, or distance, z.

This variation of the amplitude may be represented mathematically by the periodically varying function, u,

$$u = u_o \cos(kz - \omega t + \alpha) \quad \text{or} \quad u = u_o \, e^{i\alpha} e^{i(kz - \omega t)} \qquad (3.1)$$

where u_o is the maximum amplitude, ω is the angular frequency giving the rate of change of the phase with time, and k is the wave number giving the rate of change of phase with distance. For the wave oscillation of period, T, and wavelength, λ, we have,

ωt: phase change with time, $\omega = 2\pi/T$,
kz: phase change with distance, $k = 2\pi/\lambda$,
α: arbitrary initial phase.

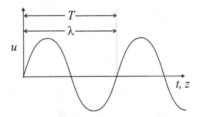

Figure 3.1. Graphical representation showing variation of wave amplitude, u, with period T, in time or λ, in space.

The wave number, k, is a scalar but is sometimes used as a vector, $\mathbf{k}$, to indicate the direction of the wave propagation along the z-axis.

3.2 Interference

The addition of amplitudes from two sources results in interference, the amplitudes may add or cancel each other depending on the relative phase. Experimentally, the interference of two light waves is well-illustrated by the famous experiment of Thomas Young in 1803. The arrangement is shown schematically in figure 3.2. Light incident on an opaque screen from the left is transmitted by two narrow parallel slits and observed at a point P.

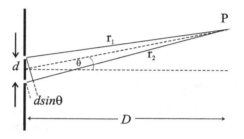

Figure 3.2. Young's slits.

For simplicity we consider the two slits to have negligible width and that they are illuminated by monochromatic plane waves of angular frequency, ω, so that the amplitude at each slit is the same, u_o. The amplitude u_p at a point P, a large distance, D, from the slits will be given by adding the amplitudes of the waves from each of the slits,

$$u_p = \frac{u_o}{r_1}e^{i(kr_1-\omega t)} + \frac{u_o}{r_2}e^{i(kr_2-\omega t)}. \tag{3.2}$$

The factors $1/r_1$ and $1/r_2$ account for the fall-off in *amplitude* according to the inverse square law for the *intensity* of the wave. Putting the path difference from each slit to the point P, $(r_1 - r_2) = d \sin \theta$, with $r_1 \sim r_2 = r$, the intensity, I_p, is found by multiplying the amplitude by its complex conjugate u_p^*,

$$I_p = 4\left(\frac{u_o}{r}\right)^2 \cos^2\left(\frac{1}{2}kd \sin \theta\right). \tag{3.3}$$

3.3 Phasors

Phasors are a convenient way to represent the amplitude and phase of a wave. The amplitude of the wave is represented by the length of a 'vector' on an Argand diagram. The phase of the wave is represented by the angle of the vector relative to the real axis of the Argand diagram. So that, as the phase changes with time or

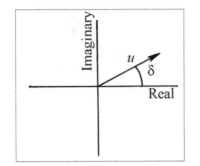

Figure 3.3. Phasor diagram.

distance, the direction of the phasor will rotate in the plane of the real and imaginary axes.

The **phasor** is then $ue^{i\delta}$ as shown in figure 3.3.

As an example of how to use phasors we consider the problem of calculating the amplitude and intensity of the interference pattern in the Young's slits experiment. Since it is the *relative* phase of the two interfering waves that matters we can arbitrarily assume the phasor representing one of the waves lies along the real axis. The other wave will have a relative phase represented by the angle, δ, as shown in figure 3.4.

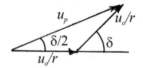

Figure 3.4. Phasor diagram for the two slit problem.

The amplitude from each slit on the screen (assumed to be approximately equal) is u_o/r.

The phase difference, δ, owing to the path difference, $d \sin \theta$, is $\delta = kd \sin \theta$.

The resultant amplitude is then,

$$u_p = 2\frac{u_o}{r} \cos(\delta/2). \tag{3.4}$$

The intensity is therefore,

$$I_p = 4\left(\frac{u_o}{r}\right)^2 \cos^2\left(\frac{1}{2}kd \sin \theta\right). \tag{3.5}$$

3.4 Diffraction from a finite slit

Most slits are not infinitesimally narrow; if they were they would not transmit much light and it would be difficult to see any interference patterns! In practice then we need to use finite slits and so we cannot ignore the effect that this has on the diffraction. We consider a monochromatic plane wave incident from the left on an aperture of width, a. The diffracted light is observed in a plane at large distance, D, from the aperture, as shown in figure 3.5. The amplitude in the plane of aperture is u_o per unit length.

An infinitesimal element of length, dy, at position, y, contributes at P an amplitude,

$$\frac{u_o dy}{r} e^{i\delta(y)}.$$

Once again the $1/r$ factor accounts for the reduction in amplitude described by the inverse square law of intensity. The phase factor is,

$$\delta(y) = k(r \pm y \sin \theta).$$

The total amplitude at P arising from all contributions across the aperture is then found by summing up all the infinitesimal contributions with amplitude and phase given by this expression. This sum is found simply by integrating the infinitesimal contributions over the aperture from $-a/2$ to $+a/2$.

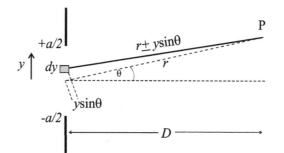

Figure 3.5. Contributions to amplitude at P from elements dy in slit.

$$u_p = \frac{u_o}{r} e^{ikr} \int_{-a/2}^{a/2} e^{ik \sin \theta \cdot y} dy \tag{3.6}$$

$$u_p = \frac{u_o}{r} \text{sinc} \, \beta \tag{3.7}$$

where $\beta = \frac{1}{2} ka \sin \theta$.

The intensity is then,

$$I_p = I(0)\text{sinc}^2 \, \beta. \tag{3.8}$$

This intensity pattern is shown as a function of β, i.e. of angle, θ, to the axis in figure 3.6

The first minimum is at $\beta = \pi$, i.e. $2\pi = \frac{2\pi}{\lambda} a \sin \theta$.

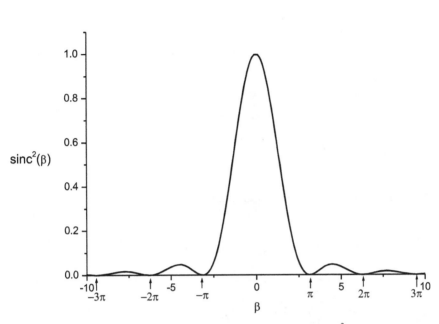

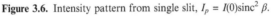

Figure 3.6. Intensity pattern from single slit, $I_p = I(0)\text{sinc}^2 \beta$.

Hence the angular width, $\Delta\theta$, of the diffraction peak, defined as the angular separation of the first minima on either side of the central peak is,

$$\Delta\theta = \frac{\lambda}{a}. \tag{3.9}$$

This simple equation represents the basic physics of all wave diffraction phenomena. The diffraction angle, $\Delta\theta$, i.e. the degree to which a wave spreads out after passing through an aperture, is proportional to the wavelength of the wave and inversely proportional to the width of the aperture. So longer waves diffract, i.e spread out more, and narrow apertures cause more spreading out than wide ones.

3.5 Diffraction from a finite slit: *phasor treatment*

We can repeat the above calculation using phasors to further illustrate how useful they are in solving diffraction problems. We represent the contributions of each of the infinitesimal elements in the slit by phasors. Figure 3.7 shows the first and last of these phasors contributing to the amplitude at P at distance, D, from the slit.

The amplitude at any given angle to the axis will be found by adding all the phasors across the slit. We represent this in the phasor diagram by adding a set of equal-length phasors with a small phase angle between successive elements. On axis, $\theta = 0$, the phasor elements are all in phase (the phase angle is zero) and add in a straight line to give a resultant, R_p, as shown in figure 3.8(a). Off axis, $\theta \neq 0$ and successive phase shifts (represented by a small angle) between adjacent phasors bend the phasor sum to form a section of a regular polygon as shown.

The phase difference between first and last phasors for $\theta \neq 0$ is,

$$\delta = ka \sin \theta.$$

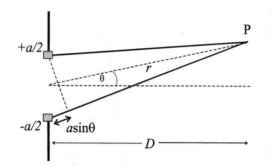

Figure 3.7. Construction showing elements at extreme edges of the aperture contributing the first and last phasors.

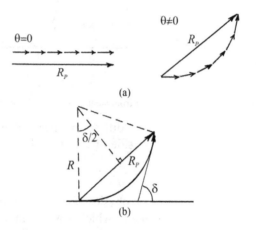

Figure 3.8. (a) Phasor diagram for a finite slit and resultant, R_p, for $\theta = 0$ and $\theta \neq 0$. (b) Phasor diagram in the limit as phasor elements tend to zero.

In the limit as the length of the phasor elements tends to zero, the phasors form an arc of a circle of radius, R. The length of the arc is R_o and the length of the chord representing the resultant is R_p, as shown in figure 3.8(b).

The amplitude at some angle, θ, relative to the amplitude at $\theta = 0$ is then,

$$\frac{\text{length of chord}}{\text{length of arc}} = \frac{2R \sin(\delta/2)}{R \cdot \delta} = \text{sinc}(\delta/2).$$

Then the intensity at θ is,

$$I(\theta) = I(0) \, \text{sinc}^2(\delta/2) = I(0)\text{sinc}^2 \beta. \qquad (3.10)$$

The first minimum occurs when the phasor arc bends to become a full circle, i.e. the phase difference between first and last phasor elements $\delta = 2\pi$ as shown in figure 3.9. As before, the angular width from the central peak to the first minimum is,

$$\theta = \frac{\lambda}{a}. \qquad (3.11)$$

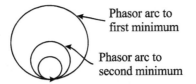

Figure 3.9. Phasor diagram showing minima for increasing phase shift, δ, between extreme edges of the slit as θ increases.

3.6 Diffraction in two dimensions

One can recall that the amplitude resulting from a plane wave illumination of an aperture of the form of a slit of width, a, in the y-direction is,

$$u_p = \frac{u_o}{r} e^{ikr} \int_{-a/2}^{a/2} e^{ik \sin \theta \cdot y} dy. \tag{3.12}$$

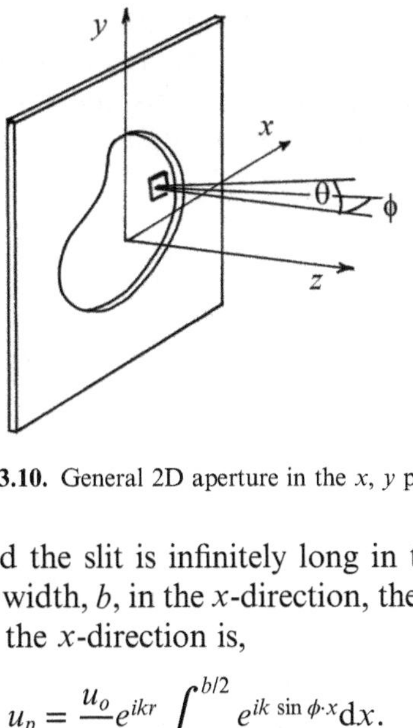

Figure 3.10. General 2D aperture in the x, y plane.

We have implicitly assumed the slit is infinitely long in the x-direction. Consider, now, the aperture to have a width, b, in the x-direction, then the angular variation of the diffracted amplitude in the x-direction is,

$$u_p = \frac{u_o}{r} e^{ikr} \int_{-b/2}^{b/2} e^{ik \sin \phi \cdot x} dx.$$

A more general aperture, shown in figure 3.10, is finite in both the x- and y-directions. So we need to consider diffraction in two dimensions. In 2D we have,

$$u_p \propto e^{ikr} \int_{-b/2}^{b/2} \int_{-a/2}^{a/2} u(x, \ y) \cdot e^{ik(\sin \phi \cdot x + \sin \theta \cdot y)} dx dy. \tag{3.13}$$

The amplitude distribution function for the aperture may be written as $u(x, y)$. For a circular aperture of diameter, a, the diffraction pattern is a circular Bessel function. In this case the angular width to the first minimum is,

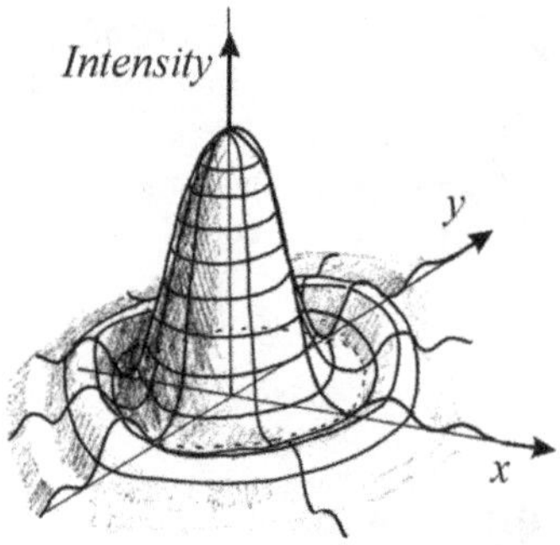

Figure 3.11. Point spread function for a circular aperture.

$$\theta = 1.22\frac{\lambda}{a}. \tag{3.14}$$

A point source imaged by a lens of focal length, f, and diameter, a, gives a pattern with a minimum of radius, $r = f\lambda/a$. This is the *point spread function, PSF*, or *instrument function* shown in figure 3.11. This function is analogous to the *impulse function* of an electrical circuit giving its response to a δ-function impulse.

Optics
The science of light
Paul Ewart

Chapter 4

Fraunhofer diffraction

So far we have considered diffraction by (a) apertures or slits illuminated by plane waves and (b) observation at a large distance where the phase difference between contributions from secondary sources in the diffracting plane separated by y is given to a good approximation by,

$$\delta = k \sin \theta \cdot y. \tag{4.1}$$

These are *special cases* where the phase difference, δ, is a *linear function* of the position, y, in the diffracting aperture. Although they are special cases they do allow a reasonably accurate treatment of diffraction in some important situations.

4.1 Fraunhofer diffraction

The common feature of the special cases just described centres on the linear variation of the phase from points across the diffracting aperture. We can therefore give a formal definition of the diffraction in these situations as follows.

> **'A diffraction pattern for which the phase of the light at the observation point is a linear function of position for all points in the diffracting aperture is *Fraunhofer diffraction*.'**

Usually the phase variation is not exactly linear. For example, the light radiating from a point source will have a spherical wave front, i.e. the points of equal phase on the wave lie on the surface of a sphere. Such a wave incident on a plane will have a phase in the plane that will not be linear. However, we can generalise our definition by giving a quantitative meaning to the term 'linear'. By *linear* we mean that the wave front deviates from a plane wave by less than $\lambda/20$ across the diffracting aperture. This situation is shown in figure 4.1 where a spherical wave arrives at, or leaves, an aperture in a plane at distance, R, from the source or observation point.

doi:10.1088/2053-2571/ab2231ch4

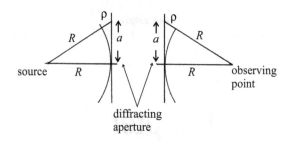

Figure 4.1. Wave-fronts incident on and exiting from a plane aperture.

The wave arriving at a point in the plane of the aperture a distance, a, from the axis will have a different phase from the axial value as a result of the path difference, ρ. From the diagram we can see,

$$(R + \rho)^2 = R^2 + a^2$$

and so for $\rho \leqslant \lambda/20$, $R \approx 10a^2/\lambda$, where we have neglected the small term, $\lambda^2/400$.

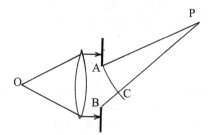

Figure 4.2. The Fraunhofer condition is satisfied by a point source at the focal length from the lens to create plane waves, the image of the source is at infinity and a plane wave is created in aperture AB.

An alternative definition of Fraunhofer diffraction that applies to any image-forming optical system may be given by the following.

'Fraunhofer diffraction is the diffraction observed in the image plane of an optical system.'

Consider a point source at the focal point of a lens as shown in figure 4.2, so that collimated light, or plane waves, are incident on an aperture behind the lens. The image of the source is at ∞. Fraunhofer diffraction will be observed at P if $BC \leqslant \lambda/20$.

The important point about this definition is that it is not necessary to specify that Fraunhofer diffraction applies only to diffraction of plane waves. Consider the case where the observation point P lies in the image plane of the lens so that curved wave-fronts converge from the lens to P. In this case, no plane waves are involved. The lens and diffracting aperture, however, can be replaced by an equivalent system where diffraction of plane waves occurs as shown in figure 4.3. Note, however, that

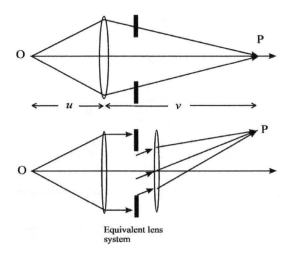

Figure 4.3. Fraunhofer diffraction observed in the image plane of a lens.

this means that plane waves are *not* necessary to observe Fraunhofer diffraction. The key criterion is that *the phase varies linearly with position in the diffracting aperture*.

A further consequence of noting that Fraunhofer diffraction is observed in the image plane is that the position of the aperture is not important—see figure 4.4, where the aperture and the lens combination may be placed at any position between the imaging lens and the image.

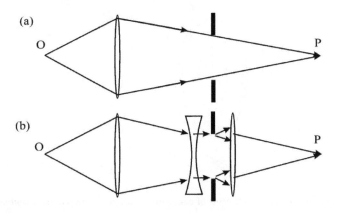

Figure 4.4. Equivalent lens system showing that Fraunhofer diffraction is independent of position of aperture. The image-forming system in (b) is equivalent to that in (a) as the source and image are in the same planes. The diffraction at the aperture in (a) is that of a curved wavefront whereas that in (b) is of a plane wave and therefore meets the Fraunhofer condition, but the systems are equivalent.

4.2 Diffraction and wave propagation

Consider a plane wave surface at $-z$ shown in figure 4.5. This reproduces itself at a second plane $z = 0$. Huygens secondary sources in the wave front radiate to a point P in the second plane.

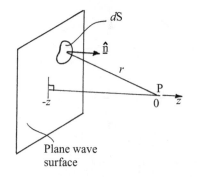

Figure 4.5. Huygens secondary sources on plane wave at $-z$ contribute to the wave at P.

The amplitude at P is the resultant of all contributions from the plane at $-z$.

$$u_p = \alpha \int \frac{u_o \, \mathrm{d}S}{r} \eta(n, r) e^{ikr} \qquad (4.2)$$

where u_o is the amplitude from element of area, $\mathrm{d}S$. The term $\eta(n, r)$ is the *obliquity factor*, which accounts for the fact that the wave propagates only in the forward direction. $\hat{\mathbf{n}}$ is a unit vector normal to the wave front and α is a proportionality constant that is to be determined.

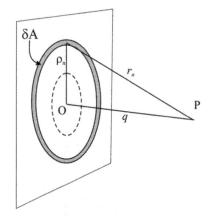

Figure 4.6. Construction of elements of equal area δA on a plane wave front centred at O.

We determine α by a self-consistency argument based on the condition that a plane wave at $-z$ must reproduce itself at $z = 0$. We consider the amplitude at a point P a distance, q, from the wave such that $q = m\lambda$, where m is an integer and $m \gg 1$, i.e. P is a large distance from O, a point on the wave front lying on a normal through P. We construct elements of the wave front of equal area, δA, centred on O as shown in figure 4.6.

The first element is a circle, the nth is an annulus of outer radius ρ_{n+1},

$$\pi\left(\rho_{n+1}^2 - \rho_n^2\right) = \delta A.$$

Consequently the difference in distance, δr, from successive elements to P is constant,

$$\delta r = r_{n+1} - r_n \cong \frac{\delta A}{2\pi r_n} \cong \frac{\delta A}{2\pi q}. \tag{4.3}$$

Therefore, the phase difference, $\delta\phi$, between waves from successive elements is also constant,

$$\delta\varphi = \frac{2\pi}{\lambda}\delta r$$

$$\delta\varphi = \frac{\delta A}{\lambda q}. \tag{4.4}$$

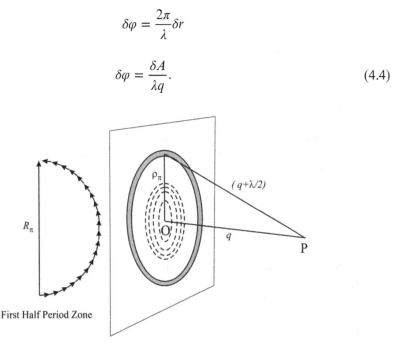

Figure 4.7. Construction of the first half period zone.

Hence we may treat contributions from each element of the wave front as a phasor.

(Note: we ignore, for the moment, the small difference in amplitude at P between successive elements arising from the small increase in distance, r_n, as ρ_n increases. We also ignore the small change in $\eta(n, r)$). The phasor diagram is shown in figure 4.7, where we see how we can add contributions of all the elements (phasors) until the last phasor added is π out of phase with the first. The area of the wave front covered by these elements is the first half period zone (1st HPZ).

The difference in path length from the outer element of the 1st HPZ to P and from O to P is $\lambda/2$, as shown in figure 4.8.

From the geometry of figure 4.7, we find by neglecting the small term $\lambda^2/4$, we find the radius of the 1st HPZ, ρ_π is given by,

$$\rho_\pi^2 = \lambda q.$$

Recalling our diffraction integral, we write, using equation (4.2), the contribution to the amplitude at P from the 1st HPZ,

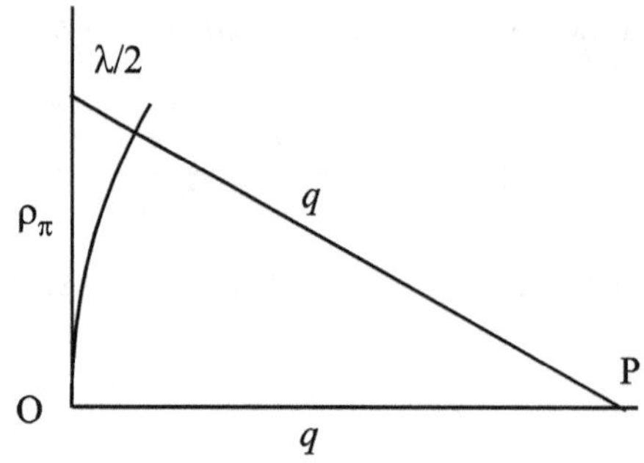

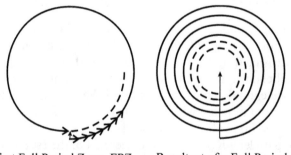

Figure 4.8. Phase shift of $\lambda/2$ arises at the edge of the 1st HPZ.

$$\alpha\frac{u_o\pi\rho_\pi^2}{q} = \alpha u_o\pi\lambda. \tag{4.5}$$

From the phasor diagram, the amplitude from the 1st HPZ is the length of the phasor arc, $\alpha u_o\pi\lambda$. The resultant, R_π, is then the diameter of the circle of which the phasor arc defines half the circumference,

$$(1/2)\pi R_\pi = \alpha u_o\pi\lambda.$$

The resultant phasor lies along the imaginary axis so,

$$R_\pi = 2i\alpha u_o\lambda.$$

We continue to add further elements until the final phasor is in phase with the first, i.e. a phase difference of 2π. The area of the wave front now defines the first full period zone, 1st FPZ. The resultant from the 1st FPZ is not exactly zero owing to the term $1/r$ (inverse square law for intensity) and the obliquity factor $\eta(n,\ r)$. Adding further elements gives a slow spiral as shown in figure 4.9.

First Full Period Zone, FPZ   Resultant of n Full Period Zones

Figure 4.9. As $n \rightarrow \infty$ the resultant of all the zones tends to half the resultant of the 1st HPZ.

Adding contributions from the whole wave (integrating over an infinite surface) gives a resultant equal to 1/2 the 1st HPZ. Therefore,

$$R_\infty = i\alpha u_o\lambda.$$

Self-consistency demands that this wave at P matches the original wave at O,

$$u_o = i\alpha u_o \lambda$$

$$\therefore \alpha = -\frac{i}{\lambda}. \qquad (4.6)$$

Hence,

$$u_p = -\frac{i}{\lambda} \int \frac{u_o dS}{r} \eta(n,\, r) e^{ikr}. \qquad (4.7)$$

This is the Fresnel–Kirchhoff diffraction integral.

Optics
The science of light
Paul Ewart

Chapter 5

Fourier methods in optics

So far we have looked at the physics underlying the phenomenon of diffraction and have seen that it is really just a manifestation of wave interference. We have seen how to calculate the diffraction patterns of some simple apertures. We now introduce a powerful mathematical method, the Fourier transform, which will be of general usefulness, especially in calculating diffraction from more complicated apertures or systems of apertures.

5.1 The Fresnel–Kirchhoff integral as a Fourier transform

The Fresnel–Kirchhoff diffraction integral tells us how to calculate the field, u_p, in an observation plane using the amplitude distribution u_o in some initial plane,

$$u_p = -\frac{i}{\lambda} \int \frac{u_o \, \mathrm{d}S}{r} \eta(\underline{n},\ \underline{r}) e^{ikr}, \tag{5.1}$$

where the limits of integration will be defined by the boundary of the aperture. We can simplify this integral by making the following approximations,

- Ignore the obliquity factor, i.e. put $\eta(\underline{n},\ \underline{r}) = 1$.
- Restrict to one dimension, $\mathrm{d}S \to \mathrm{d}x$.
- Ignore the $\frac{1}{r}$ term by considering only a small range of r, in other words, when r doesn't vary very much and is approximately a constant, any small variation will not change the relative amplitude of the contributions from different parts of the aperture.
- Use the Fraunhofer condition, $e^{ikr} = e^{ikr'} e^{ik \sin \theta x}$.
- Absorb the $e^{ikr'}$ term into the constant of proportionality.

Since the integral will be zero wherever the amplitude function, $u(x)$, is zero, the limits of integration can safely be extended to infinity. The amplitude, u_p, as a function of angle, θ, is then, with $\beta = k \sin \theta$,

$$u_p \Rightarrow A(\beta) = \alpha \int_{-\infty}^{\infty} u(x)e^{i\beta x}\mathrm{d}x. \tag{5.2}$$

This expression for $A(\beta)$ has the form of the Fourier transform of $u(x)$.

So we have an important result.

The Fraunhofer diffraction pattern is the Fourier transform of the amplitude function of the diffracting aperture.

More precisely, the Fraunhofer diffraction pattern, expressed as the amplitude as a function of angle, is the Fourier transform of the function representing the amplitude of the incident wave, as a function of position, in the diffracting aperture. The Fraunhofer diffraction is expressed as a function of $\beta = k \sin \theta$, where θ is the angle of the diffracted wave relative to the wave vector, $\underline{k}$, of the wave incident on the aperture.

We can see that the Fourier transform provides a straightforward way to calculate the diffraction pattern of any given aperture.

The inverse transform relation is,

$$u(x) = \frac{1}{\alpha} \int_{-\infty}^{\infty} A(\beta)e^{-i\beta x}\mathrm{d}\beta. \tag{5.3}$$

The interesting thing to note is that if we can measure the diffraction pattern, and thus determine the function $A(\beta)$, we can use this to reconstruct the amplitude distribution in the original aperture. This process is the basis for analysing x-ray diffraction patterns and using them to construct the shapes of the diffracting structures such as crystals or large molecules. This was, essentially, the way the double-helix structure of DNA was established.

5.2 The convolution theorem

The convolution of two functions, $f(x)$, and, $g(x)$, is a new function, $h(x)$, defined by,

$$h(x) = f(x) \otimes g(x) = \int_{-\infty}^{\infty} f(x')g(x - x')\mathrm{d}x'. \tag{5.4}$$

The Fourier transforms, F.T., of each of the functions may be denoted by the following,

$$\text{F.T. of } f(x) \text{ is } F(\beta)$$
$$\text{F.T. of } g(x) \text{ is } G(\beta)$$
$$\text{F.T. of } h(x) \text{ is } H(\beta).$$

The convolution theorem states that the Fourier transform of a convolution of two functions is the product of the Fourier transforms of each of the two functions,

$$H(\beta) = F(\beta) \cdot G(\beta). \tag{5.5}$$

The value of this concept of a convolution is that it turns out that we can represent some complicated functions as convolutions of more simple functions. Then, if we know, or can calculate, the Fourier transform of the simpler functions, we can use the convolution theorem to find the Fourier transforms of the more complicated function. This will then allow us to calculate the diffraction patterns of complicated apertures. We will illustrate below how this works by considering some examples.

5.3 Some useful Fourier transforms and convolutions

Firstly, we will introduce some simple functions and their Fourier transforms. Then we will give some examples of aperture functions that can be represented by convolutions of simpler functions.

(a) We can represent a wave of constant frequency, β_o, as a function of time, t,

$$v(t) = V_o e^{-i\beta_o t}$$

$$\text{F.T. } \{v(t)\} = V(\beta) = V_o\delta(\beta - \beta_o), \tag{5.6}$$

i.e. $V(\beta)$ represents the spectrum of a monochromatic wave of frequency, β_o, and is a delta function in frequency space, as shown in figure 5.1.

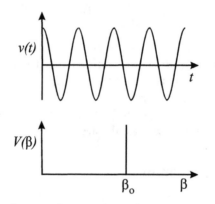

Figure 5.1. A wave of constant frequency (monochromatic) and its Fourier transform.

Alternatively, the inverse transform relations allow us to represent the F.T. of a delta function, e.g. a delta function in time, $v'(t)$, will be transformed to a function of frequency

$$v'(t) = V_o\delta(t - t_o) \text{ because the inverse F.T. } \{v(t)\} = V(\beta) = V_o e^{-i\beta t_o} \tag{5.7}$$

(b) The double slit function, i.e. two delta functions separated by d,

$$v_b(x) = \delta\left(x \pm \frac{d}{2}\right)$$

$$V_b(\beta) = 2\cos\left(\frac{1}{2}\beta d\right) \tag{5.8}$$

The phase factor in equation (5.8) can be written as a function of angle as $\beta = k \sin \theta$. We recognise this result, equation (5.8), as the same form as we found for the distribution of amplitude in a two-beam interference pattern, equation (3.4),

$$u_p = 2\frac{u_o}{r} \cos(\delta/2) \qquad (3.4)$$

where $\delta = kd \sin \theta$ or $\delta = \beta d$.

Alternatively, the inverse transform relations allow us to represent the F.T. of a delta function, e.g. a delta function in time $v'(t)$ will be transformed to a function of frequency.

$$v(t) = V_o\delta(t - t_o) \text{ because the inverse F.T. } \{v(t)\} = V(\beta) = V_o e^{-i\beta t_o}. \qquad (5.9)$$

(c) A comb of delta functions

A comb of delta functions, shown in figure 5.2, may be represented by the function,

$$v_c(x) = \sum_{m=0}^{N-1} \delta(x - mx_s).$$

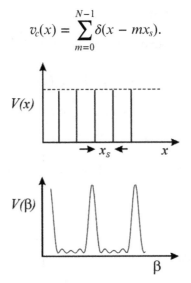

Figure 5.2. A comb of δ-functions and their transform.

The F.T. of $v_c(x)$ is then,

$$V_c(\beta) = e^{i\alpha} \frac{\sin\left(\frac{1}{2}N\beta x_s\right)}{\sin\left(\frac{1}{2}\beta x_s\right)} \qquad (5.10)$$

where,

$$\alpha = \frac{1}{2}(N - 1)\beta x_s.$$

The factor $e^{i\alpha}$ in equation (5.10) is simply the consequence of starting our comb at $x = 0$. This factor can be eliminated by shifting our comb to sit symmetrically about the origin. This result illustrates the 'shift theorem'.

(d) The top-hat function:

$$v_d(x) = 1 \text{ for } |x| < \frac{a}{2}$$

$$v_d(x) = 0 \text{ for } |x| > \frac{a}{2} \tag{5.11}$$

$$V_d(\beta) = a \, \text{sinc}\left(\frac{1}{2}\beta a\right).$$

This result, again, we recognise as the result of a previous calculation by direct addition of the amplitudes or by using phasors. Equation (5.11) is equivalent to the form of the diffraction pattern of a single slit of width, a (see equations (3.7), (3.8) and (3.10)).

Now some useful convolutions.

(e) The double slit—a convolution of two delta (slit) functions with a top-hat function,

$$v_s(x) = v_b(x) \otimes v_d(x).$$

(f) The grating function—a convolution of a comb with a top hat,

$$v_g(x) = v_c(x) \otimes v_d(x).$$

(g) The triangle function—a convolution of a top-hat function with another identical top-hat function,

$$v_\Delta(x) = v_d(x) \otimes v_d(x).$$

This is a self-convolution. The self-convolution is known also as the *autocorrelation function*.

5.4 Fourier analysis

A periodic function, $V(t)$, may be represented by a Fourier series,

$$V(t) = c_o + \sum_{p=1}^{\infty} c_p \cos(p\omega_o t) + \sum_{p=1}^{\infty} s_p \sin(p\omega_o t). \tag{5.12}$$

$V(t)$ is the result of *synthesis* of the set of Fourier components. Fourier *analysis* is the reverse process, i.e. finding the components (amplitude and phase) that make up $V(t)$. The coefficients are found by multiplying the function by $\cos(q\omega_o t)$ or $\sin(q\omega_o t)$ and integrating the function over a period, τ, of the oscillation. All the resulting terms in the sum vanish except when $q = p$. Thus we find the coefficients, provided we have an analytic expression for $V(t)$ and can integrate it!

$$s_p = \frac{2}{\tau} \int_0^\tau V(t)\sin(p\omega_o t)\mathrm{d}t$$

$$c_p = \frac{2}{\tau} \int_0^\tau V(t)\cos(p\omega_o t)\mathrm{d}t$$

$$c_o = \frac{1}{\tau} \int_0^\tau V(t)\mathrm{d}t.$$

In general,

$$V(t) = \sum_{p=1}^\infty A_p e^{-ip\omega_o t}$$

$$A_p = \frac{1}{\tau} \int_0^\tau V(t)e^{ip\omega_o t}\mathrm{d}t.$$

(5.13)

This last expression represents a Fourier transform—suggesting that this operation analyses the function $V(t)$ into its Fourier components and gives their amplitudes, A_p.

5.5 Spatial frequencies

Consider a plane wave falling normally on an infinite screen with amplitude transmission function,

$$u(x) = 1 + \sin(\omega_s x), \tag{5.14}$$

i.e. a grating with a periodic pattern of width,

$$d = \frac{2\pi}{\omega_s}.$$

This defines the spatial frequency,

$$\omega_s = \frac{2\pi}{d}. \tag{5.15}$$

The Fraunhofer diffraction pattern is then,

$$A(\beta) = \alpha \int_{-\infty}^\infty u(x)e^{i\beta x}\mathrm{d}x \tag{5.16}$$

where $\beta = k \sin \theta$.
    We find,

$$A(\beta) = 0, \text{ except for } \beta = 0, \pm \omega_s,$$

i.e.

$$\sin \theta \approx \theta = 0 \text{ or } \pm \frac{\lambda}{d}. \tag{5.17}$$

The sinusoidal grating has a Fraunhofer diffraction pattern consisting of zero order and first orders only at values of an angle given by $\theta = \pm\lambda/d = \pm\lambda\omega_s/2\pi$.

If our sinusoidal grating had an additional spatial frequency, ω_n, in the transmission pattern, this additional grating will lead to additional first orders at $\theta = \pm\lambda\omega_n/2\pi$. The presence of the two sets of first order diffraction beams thus indicates the presence of the two spatial frequencies in the amplitude function. One can then think of the angular diffraction pattern providing a way to analyse the spatial pattern in the grating. This is another way to think of the Fourier transform, which produces the form of the angular pattern, as an operation to analyse the spatial structure in the diffracting aperture.

(Note: a finite screen will result in each order being spread by the diffraction pattern of the finite aperture, i.e. the 'spread function' of the aperture.)

5.6 Abbé theory of imaging

We consider an object consisting of an infinite screen having a sinusoidal transmission described by a function, $u(x)$, so that the amplitude transmission repeats with a spacing, d. This acts as an object at a distance, u, from a lens of focal length, f.

Diffraction orders are waves with parallel wave vectors at angles $\theta = 0$ and $\theta = \pm\lambda/d$.

A lens brings these parallel waves to a focus as 'points' in the focal plane separated by $a = f\lambda/d$. Apart from a phase factor, the amplitude in the focal plane is the F.T. of $u(x)$. This plane is the *Fourier plane*. The amplitude distribution in this plane is the Fourier transform of that in the object plane, $u(x)$.

The zero and first order 'points' act as coherent sources giving two-beam interference at positions beyond the focal plane as shown in figure 5.3. For a finite grating the 'points' will also be spread by diffraction at the effective aperture of the grating. (Note that we can describe such a grating as a convolution of an infinite sine wave with a top-hat function.)

In the image plane, distance, v, from the lens, the interference pattern is maximally sharp, $v = f + D$. The interference pattern, i.e. the image is a sinusoidal fringe system with spacing,

$$d' = \frac{D\lambda}{a}. \tag{5.18}$$

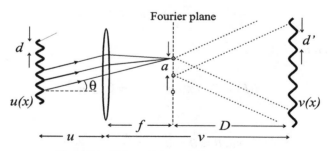

Figure 5.3. Object $u(x)$ imaged by a lens to $v(x)$.

From the geometry, we see,

$$\frac{d}{u} = \frac{d'}{v}.$$

Hence,

$$\frac{1}{u} + \frac{1}{v} = \frac{1}{f}. \tag{5.19}$$

This is just the 'thin lens equation' relating the object and image distances, u and v, respectively, for the imaging of an object by a lens of focal length f (see equation (2.2)).

Thus, we have explained the process of image formation by two sequential processes, each of which can be described by a Fourier transform. First, the diffraction at the object plane leads to a pattern in the Fourier plane, described by the Fourier transform of the object distribution, $u(x)$. This pattern in the Fourier plane acts as a new object distribution in the form of a set of coherent sources. Second, the diffraction of the beams from these coherent 'sources', the zero and first order diffraction spots, leads to the interference pattern that reproduces the original amplitude distribution. This diffraction is described by another Fourier transform to produce the image. Thus the image is formed by the Fourier transform of a Fourier transform which recovers the original form of the object distribution except for a magnification factor, M,

$$M = \frac{d'}{d} = \frac{v}{u}. \tag{5.20}$$

From equation (5.16) that defines the Fourier transform we can see that two successive transforms, each with a factor α, will produce a factor of $(-i)^2 = -1$ (see equation (4.6)). The image will therefore be inverted with respect to the original amplitude distribution.

Any object amplitude distribution may be synthesised by a set of sinusoidal functions. Each Fourier component with a specific spatial frequency contributes $\pm$ orders to the diffraction pattern at specific angles, θ, to the axis. The aperture, a, of the lens and object distance, u, determine the maximum angle, θ_{max}, from which light may be collected. Diffraction orders at angles greater than θ_{max} do not contribute to the final image. The corresponding spatial frequencies will be missing from the image. Higher spatial frequencies contribute to sharp edges in the object distribution. The lack of high spatial frequencies in the image leads to blurring and loss of resolution.

We note that the discussion so far is valid only for *coherent* light, i.e. light waves having a fixed phase relationship across the aperture in the object plane. In practice, for microscopic objects, this condition is quite well fulfilled even for white light illumination because the phase will be practically the same across the small range of the wavefront encountering a microscopic object.

5.7 Spatial resolution of the compound microscope

Diffraction will ultimately limit the resolution of images formed by any lens system. As an example we consider the compound microscope. Figure 2.8 shows the arrangement of the compound microscope in which a very short focal length lens, the objective, forms a real, inverted image of the specimen in the image plane, giving a linear magnification of v/u. The eyepiece acts as a simple magnifier used to view the real image that is located in the front focal plane of the eyepiece giving a virtual image at infinity. This allows viewing with minimum eyestrain. The minimum dimension of spatial structure in the object, d_{min}, that can be resolved is such that the associated diffraction order will be at the maximum angle, θ_{max}, that can be collected by the objective lens,

$$\sin \theta_{max} = \lambda / d_{min}. \tag{5.21}$$

Spatial frequencies corresponding to structures with dimensions smaller than d_{min} will result in diffraction to larger angles (see figure 5.4). This diffracted light will miss the objective and thus not appear in the image. These small structures will therefore not be resolved in the image.

The discussion so far has assumed that the medium between the object and objective is air with refractive index, $n_{air} = 1$. If, however, the objective and object is immersed in oil of refractive index, n_o, the spatial resolution can be increased. In the case of the oil immersion microscope, equation (5.21) is modified,

$$n_o \sin \theta_{max} = \frac{\lambda}{d_{min}}.$$

Since $n_o > 1$, for the same objective lens diameter as before, the value of d_{min} will be smaller and the spatial resolution is increased. $n_o \sin \theta_{max}$ is the numerical aperture and defines the ultimate spatial resolution of the device.

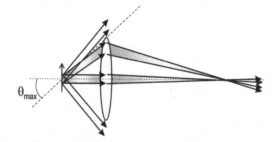

Figure 5.4. First order diffracted waves from spatial structures $> d_{min}$ are collected by the lens and interfere with zero order waves in the image plane to form sinusoidal structure in the image. Light from smaller spatial structures (higher spatial frequencies) are diffracted to angles greater than θ_{max}, miss the objective and do not interfere with zero order to contribute to the image.

5.8 Diffraction effects on image brightness

Normal image brightness is determined by the *f/no.* of the optical system i.e. f/d_A where d_A is the limiting aperture. The brightness of the image will then be proportional to the square of the aperture diameter. When the image size approaches the order of the point spread function, i.e. has an angular size $\sim \lambda/d_A$, light is lost from the image by diffraction. This is diffraction limited imaging. Increasing the diameter of the aperture under these conditions will have two effects. Firstly, the brightness is increased as a result of the increased area transmitting the light. Secondly, the diffraction angle is reduced, i.e. the point spread is reduced and so the light is concentrated into an image of smaller area. Both these effects are proportional to the square of the aperture diameter. Summarising, we find:

For non-diffraction limited imaging:

$$\text{Image brightness} \propto d_A^2.$$

For diffraction limited imaging:

$$\text{Image brightness} \propto d_A^4.$$

Chapter 6

Optical instruments and fringe localisation

Optical instruments for spectroscopy use interference to produce a wavelength-dependent pattern. Measurement of these patterns, or fringes, allows us to infer the spectral content of the light. The interfering beams are produced either by *division of wavefront* or by *division of amplitude*. In the Young's slit arrangement the wavefront is divided into two components by transmission through the slits in an opaque screen. In a similar way the diffraction grating divides the wavefront into multiple beams. We will consider also two instruments in which the amplitude is divided; the Michelson interferometer divides the amplitude into two beams and the Fabry–Pérot interferometer divides the amplitude into multiple beams. It is important to know where to look for the fringes so that they can be reliably measured. So first we consider the general question of fringe localisation.

6.1 Division of wavefront

6.1.1 Two-slit interference, Young's slits

If the two slits in the screen, illuminated from the left, shown in figure 6.1 were effectively point sources, the waves spreading out on the far side would overlap and interfere everywhere. The fringes are *non-localised*. With finite slits, however, the fringes are usually visible only where the diffracted beams overlap with sufficient intensity. Usually the observation screen is placed at a large enough distance that the fringes are observed under the Fraunhofer condition—the interfering waves will be effectively plane waves. In the region closer to the slits an interference pattern will be observed known as *Fresnel interference* where the pattern is affected by the waves having curved wavefronts.

6.1.2 *N*-slit diffraction, the diffraction grating

Figure 6.2 shows an opaque screen containing a large number, N, of parallel, narrow slits illuminated from the left by monochromatic plane waves. Light diffracted at the

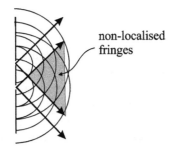

Figure 6.1. Young's slit fringes are observed throughout the region beyond the screen containing the two slits.

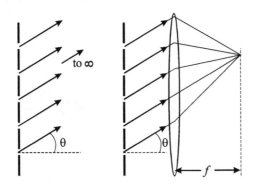

Figure 6.2. Diffraction grating fringes, localised at infinity or in the focal plane of a lens.

same angle to the axis, θ, will interfere at infinity and so this is where the fringes will be localised. These parallel rays will be brought together in the focal plane of a lens and thus allow the pattern at infinity to be viewed on a screen. In both cases, the fringes are fringes of *equal inclination localised at infinity*, or in the image plane of the lens system, and in both cases the interference meets the condition for Fraunhofer diffraction.

6.2 Division of amplitude

We will describe the details of the operation of two important spectroscopic instruments—the Michelson and Fabry–Pérot interferometers—later in the book. The essential physics, however, of both instruments may be described by considering the interference patterns produced in the following arrangements that divide the amplitude of the wave into separate beams, which are subsequently re-combined or overlapped to produce interference. The interference may involve two beams (Michelson) or multiple beams (Fabry–Pérot). The situations are modelled by the reflection of light from a source at two surfaces. The source may be a point or extended and the surfaces may be at an angle (wedged) or parallel. The images of the source in the reflecting surfaces act as two effective sources.

6.2.1 Point source

(a) Wedge

Monochromatic waves emanating from a point source, O, and reflected at the two surfaces of a wedge, as shown in figure 6.3, will appear to an observer to be coming apparently from two points P and P′. This system is equivalent to 2-point sources or a Young's slit situation. The eye viewing the two apparent sources through the wedge will see a constructive interference fringe at a position in the wedge where the thickness corresponds to an integral number of wavelengths. Therefore, the fringes are *non-localised* fringes of *equal thickness* and will appear as straight line fringes close to the apex of the wedge. Further away from the apex of the wedge the fringes will become curved owing to the spherical nature of the wavefronts.

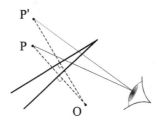

Figure 6.3. A point source, O, provides images P, P′ in reflecting surfaces forming a wedge.

(b) Parallel

A monochromatic point source reflected at two parallel surfaces is similar to the wedge situation and provides, effectively, 2-point sources, as shown in figure 6.4. Again an interference fringe is caused when the path difference of light from the two 'sources' is an integer number of wavelengths. As the angle of inclination of the reflected rays at the surface changes, a change in the path difference from the two sources will also occur leading to a set of fringes at different positions corresponding to the different angles of incidence. These fringes are *non-localised* fringes of *equal inclination.*

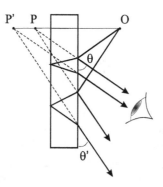

Figure 6.4. A point source reflected in two parallel surfaces again provides two images P, P′ that act as apparent sources of light waves.

6.2.2 Extended source

When an extended source is used the situation is changed as each point in the source will provide a set of independent interference fringes. The fringe patterns are independent because, for most sources of light, there is no fixed phase relationship between the waves from different parts of the source. These extended sources have no spatial coherence and so the waves from different parts will not produce stable interference patterns. Nonetheless, a monochromatic extended source will still produce visible interference patterns if certain conditions are satisfied. An atomic vapour lamp, such as a sodium lamp used for street lighting, will provide an extended source of monochromatic radiation. We can consider what happens to the light from different points in an extended source when it is reflected from two surfaces of a wedge or of a parallel plate.

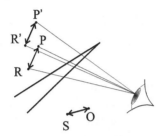

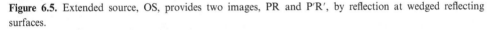

Figure 6.5. Extended source, OS, provides two images, PR and P′R′, by reflection at wedged reflecting surfaces.

(a) Wedge

Figure 6.5 shows a source extended along a line OS. The point O produces two reflected images in each face of a wedge leading to two apparent sources P and P′. As explained above, these two 'sources' will produce non-localised fringes as in the Young's slit situation. However, every point on the extended source produces its own set of non-localised fringes. Overlap of all these patterns, corresponding to a wide range of path differences, results in no visible fringes. However, at the apex of the wedge, the path difference is zero or very nearly zero, and is the same for all points on the effective sources so fringes *are* visible in this region. The zero-order fringe is a straight line fringe in the plane of the wedge. Other low-order fringes may be seen close to the apex if the source is not too large and the wedge angle not too big. The fringes are of *equal thickness* and *localised* in the plane of the wedge. An example of such a pattern is seen in the phenomenon of Newton's rings when a small air wedge is formed between a convex lens and a flat surface and illuminated by a monochromatic source. A 'zero-order' circular fringe is seen at the contact point surrounded by concentric circular fringes whose spacing decreases with increasing radius as a result of the increasing thickness of the air gap.

(b) Parallel

A practically important case is that of an extended monochromatic source reflected at two parallel surfaces. We will see later that this is the situation in some widely-used

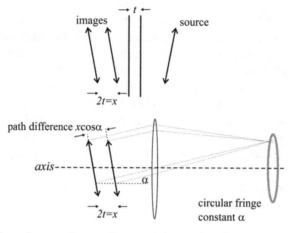

Figure 6.6. The top figure shows two images of an extended source by reflection in a parallel slab of thickness, *t*. The bottom figure shows fringes of equal inclination formed in the focal plane of a lens by light from the two images of the source.

interferometers. Figure 6.6 shows two parallel surfaces such as those of a parallel-side plate of glass of thickness, *t*. The images of any point on the extended source will be formed at positions separated by 2*t* and will act as if they were coherent sources for interference. The fringes formed by 'pairs' of point sources on the extended source will be narrow in the region close to the plate. The two-slit, or two-source, interference pattern that we calculated for the case of Young's slits will give us a measure of the fringe width as follows. From equations (3.3) and (3.5) we have,

$$I_p = 4\left(\frac{u_o}{r}\right)^2 \cos^2\left(\frac{1}{2}kd \sin\theta\right) = I_0 \cos^2\delta. \qquad (6.1)$$

The separation of the sources is *d* and the fringes will have peaks at values of the phase difference, δ, equal to integer values of π, i.e. 0, π, 2π ... $n\pi$. The angular separation of the peaks, $\Delta\theta$, is given by a change in the value of δ of π,

$$\frac{\pi}{\lambda}d \sin\Delta\theta = \pi.$$

For small angles (sin $\Delta\theta = \Delta\theta$), $\Delta\theta = \frac{\lambda}{d}$. The fringe width at distance *D* will then be,

$$\Delta y = \frac{D\lambda}{d}. \qquad (6.2)$$

Thus, for small values of *D*, the fringes formed by two points in the extended source are narrow. Another pair of points on the source displaced by a distance, *y*, will form another set of fringes displaced in the observation plane as shown in figure 6.7.

When $y \sim \Delta y/2$ the bright fringes will overlap the dark fringes leading to uniform brightness in the pattern, i.e. no fringes will be visible as a result of the overlapping of the fringe systems from different parts of the extended source. As we move away along the axis to larger distances, i.e. larger values of *D*, the fringes become wider.

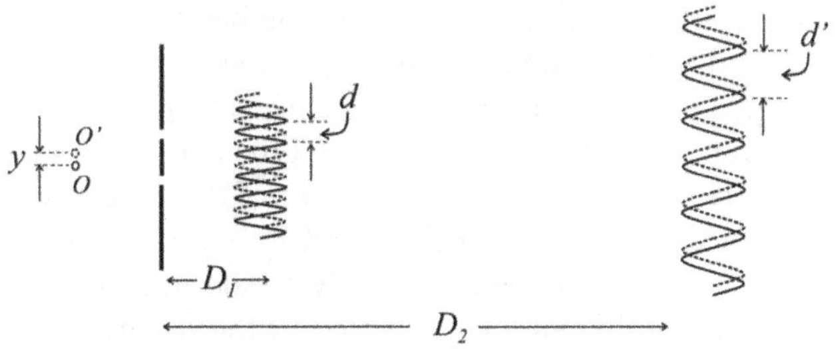

Figure 6.7. Interference fringes formed by light from two points, O and O', separated by distance, y, on an extended source. At a short distance from the slits, D_1, bright fringes from O overlap dark fringes from O' to produce uniform illumination. At large distances, $D_2 \gg D_1$, $(D \to \infty)$ the fringe width exceeds the displacement of the fringe patterns and fringes become visible (see text for details).

Eventually the fringe width becomes greater than the displacement of the fringe patterns from all the points on the extended source and then the fringes become visible. In the limit of going to infinity the displacement of the sources on the extended source becomes insignificant and the source acts effectively as a point source. The fringes that become visible are fringes of *equal inclination* and *localised at infinity*. (Note that we don't need to take the idea of infinity too literally here because, if we did, then we would have to accept that the fringes themselves are of infinite width, according to equation (6.2), when $D \to \infty$!)

These fringes are more conveniently observed in the focal plane of a lens, e.g. the eye (figure 6.6). As described above, the reflecting surfaces separated by t lead to two images separated by $2t$ or $x = 2t$. Parallel light at an angle of inclination, α, to the axis from equivalent points on the effective sources will have a path difference of $x \cos \alpha$.

The phase difference, δ, will then be,

$$\delta = \frac{2\pi}{\lambda} x \cos \alpha. \qquad (6.3)$$

The waves are brought together in the focal plane of the lens. Bright fringes (constructive interference) occur when the phase difference $\delta = p2\pi$, ($p = integer$) or,

$$x \cos \alpha = p\lambda. \qquad (6.4)$$

For small angles the angular size of the fringes is given by,

$$\alpha_p^2 - \alpha_{p+1}^2 = \frac{2\lambda}{x}.$$

The radii of these fringes in the focal plane of a lens will be r_p and r_{p+1}, such that,

$$r_p^2 - r_{p+1}^2 = \frac{2f^2\lambda}{x}, \qquad (6.5)$$

where f is the focal length of the lens.

As x increases, the circular fringes get closer together and the pattern appears to gain new fringes added from the outside. As x decreases $\rightarrow 0$ the fringe diameters get larger and fill the field of view. In this case new fringes appear at the centre of the pattern and grow outwards. Eventually, when x reaches zero the central bright fringe will fill the entire field of view. The behaviour of the fringes formed by parallel surfaces will be important for the Michelson and Fabry–Pérot interferometers.

It is worth noting at this point that generally speaking fringes will be observed only if the light source is coherent, i.e. the phase of the waves originating from the source maintain a fixed or constant phase relationship. This usually requires monochromatic light since the phase of different wavelengths or frequencies will get out of phase as the wavelength or frequency difference increases. However, when the path difference is zero, or very close to it, then all the interfering waves will have the same relative phase, i.e. zero. In this situation fringes may be observed within a narrow range around zero path difference even with 'white light'. This provides a convenient way of finding the exact position of zero path difference in some interference instruments—the device is adjusted whilst being illuminated by white light and fringes appear only when the required path difference is very close to zero.

As the descriptions above have shown, interference produces fringe patterns with shapes and sizes that depend on the wavelength, λ. These wavelength-dependent interference patterns can be used to analyse the spectrum of light by distinguishing the patterns arising from different wavelengths. We now turn to consider some practical instruments using this principle.

Optics
The science of light
Paul Ewart

Chapter 7

The diffraction grating spectrograph

A diffraction grating consists of an array of narrow, closely spaced, parallel reflecting strips or transmitting slits. To understand how a grating can be used to record the spectrum of light, i.e. how the intensity of light from a given source varies with wavelength, we will look at how the interference fringes are formed, firstly from the simplest grating consisting of only two slits and then see the effect on the pattern of adding more slits. We will use phasors to give a physical insight into how the patterns depend on the number of slits—each acting effectively as a source of light— and how to calculate the intensity distribution in the interference pattern.

7.1 Interference pattern from a diffraction grating

Consider a plane monochromatic wave of wavelength, λ, incident normally on a reflecting or transmitting grating of N slits separated by a distance, d. The amplitude contributed by each slit is u and the intensity of the interference pattern is found by adding amplitudes and taking the squared modulus of the resultant. We can, as explained in chapter 1, use phasors to represent the amplitude and phase of the wave contributed by each slit and to find the resultant.

7.1.1 Double slit, $N = 2$

It is straightforward to draw the two phasors representing the amplitude, u, from each slit as a function of the diffraction angle at some particular values of the phase difference, δ, e.g. 0, π and 2π. This allows us to plot the amplitude and the intensity as a function of δ as shown in figure 7.1. The phasor diagram is the same as figure 3.4 and equation (3.5), which gives the resultant intensity, tells us that the shape of the intensity function is a 'cosine squared'. For convenience we reproduce this function as equation (7.1)

doi:10.1088/2053-2571/ab2231ch7

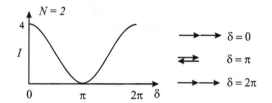

Figure 7.1. Intensity pattern and associated phasor diagram for 2-slit interference.

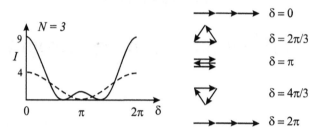

Figure 7.2. Phasor diagrams for 3-slit interference and intensity pattern.

$$I(\theta) = 4u^2 \cos^2\left(\frac{\delta}{2}\right) \tag{7.1}$$

where,

$$\delta = \frac{2\pi}{\lambda}d \sin\theta. \tag{7.2}$$

Equation (7.2) expresses the phase difference between waves propagating from adjacent slits at an angle, θ, arising from the optical path difference, $d\sin\theta$. The pattern is thus seen to be a series of $(\text{cosine})^2$ fringes with principal maxima, where both the amplitude contributions, u, are in phase, represented by the phasors pointing in the same direction giving a maximum value of $2u$. These maxima occur at $\delta = 0$, $n2\pi$, and have intensity $4u^2$. There is just one minimum between each principal maximum.

7.1.2 Triple slit, $N = 3$

Using the phasor diagrams in figure 7.2 we can find the resultant amplitude of three slits at selected values of δ between the two principal maxima at $\delta = 0$ and 2π.

(a) $\delta = 0$, $n2\pi$ Principal maxima of intensity $9u^2$.
(b) $\delta = 2\pi/3$ Minimum/zero intensity.
(c) $\delta = \pi$ Subsidiary maxima of intensity u^2.
(d) $\delta = 4\pi/3$ Minimum/zero intensity.

There are two minima between the principal maxima. The width of each principal maximum is determined by the difference in δ between the peak and the minimum on either side, $2\pi/N$ with, in this case, $N = 3$.

7.1.3 Multiple slit, $N = 4$, etc

In a similar manner we can treat the case of four slits and find principal maxima at $\delta = 0$, $n2\pi$ of intensity $16u^2$ with 3 minima between the principal maxima. A general pattern emerges where we have principal maxima at $\delta = 0$, $n2\pi$, the intensity is proportional to N^2 with $(N - 1)$ minima at phase values of $n2\pi/N$ and the width of the principal maxima, $(2 \times 2\pi/N)$, is inversely proportional to the number of slits N. For increasing values of N we will see principal maxima occurring at regular intervals, growing in intensity with N^2 and becoming sharper as the peaks narrow with increasing N. This is shown schematically in figure 7.3.

The amplitude of N phasors is,

$$A = u + ue^{i\delta} + ue^{i2\delta} + \cdots + ue^{i(N-1)\delta}. \tag{7.3}$$

This is the sum of a geometric progression with the common ratio, $e^{i\delta}$, so we find,

$$A = u\frac{(1 - e^{iN\delta})}{(1 - e^{i\delta})}.$$

Rewriting this as,

$$A = u\frac{e^{iN\delta/2}}{e^{i\delta/2}}\frac{(e^{-iN\delta/2} - e^{iN\delta/2})}{(e^{-i\delta/2} - e^{i\delta/2})}$$

and using the complex exponential form for the sin function,

$$A = ue^{i(N-1)\delta/2}\frac{\sin(N\delta/2)}{\sin(\delta/2)}.$$

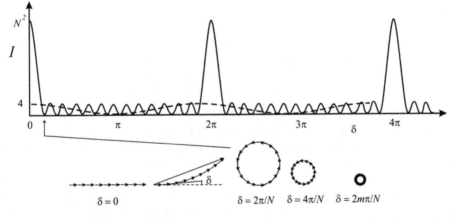

Figure 7.3. Phasor diagrams for N-slit interference and the resulting intensity pattern. As N increases the zero resultants occur when the phasors overlap the previous set forming circles of decreasing diameter—the total length of the phasors remains the same. The dashed line shows the pattern for 2-slit interference. Note that the diagram does not take account of the diffraction pattern resulting from diffraction from each slit of finite width. Principal maxima (diffraction orders) occur at phase values $n2\pi$.

The intensity is found by multiplying the amplitude by its complex conjugate, i.e. AA^*. Denoting the scaling factor by the intensity value on the axis, $I(0)$, we find,

$$I(\theta) = I(0) \frac{\sin^2\left(\frac{N\delta}{2}\right)}{\sin^2\left(\frac{\delta}{2}\right)}. \tag{7.4}$$

7.2 Effect of finite slit width

We can consider a grating of N slits of width, a, separated by d, as a convolution, $h(x)$, of a comb of δ-functions, $f(x)$, with a single slit (top-hat function), $g(x)$,

$$h(x) = f(x) \otimes g(x),$$

where,

$$f(x) = \sum_{p=0}^{N-1} \delta(x - pd) \, ;$$

$$g(x) = 1, \text{ for} \left\{-\frac{a}{2} < x < \frac{a}{2}\right\};$$

$$g(x) = 0, \text{ for} \left\{|x| > \frac{a}{2}\right\}.$$

Denoting the Fourier transforms of these functions as follows,

$$F(\beta) = F.T.\{f(x)\}, \quad G(\beta) = F.T.\{g(x)\} \text{ and } H(\beta) = F.T.\{h(x)\}.$$

We can then use the convolution theorem to write,

$$H(\beta) = F(\beta) \cdot G(\beta).$$

Hence,

$$|H(\beta)|^2 = I(\theta) = I(0) \frac{\sin^2\left(\frac{N\delta}{2}\right)}{\sin^2\left(\frac{\delta}{2}\right)} \cdot \frac{\sin^2\left(\frac{\gamma}{2}\right)}{\left(\frac{\gamma}{2}\right)^2}, \tag{7.5}$$

where $\gamma = ka \sin \theta$ and, from equation (7.2), $\delta = kd \sin \theta$.

Figure 7.4 shows the effect of the single slit diffraction on the distribution of intensity in the overall diffraction pattern. When the minimum of the single slit diffraction falls at the same position as a particular order, that order will have zero intensity. This is the reason for the so-called 'missing orders' sometimes (not) observed! The angular position of the principal peak of order, n, is set by $\theta_n \sim n\lambda/d$, where d is the slit separation. The angular position of the minimum is set by $\theta_{min} \sim \lambda/a$, where a is the slit width. If the ratio of the slit separation to width is an integer, p, i.e. $p = d/a$, then $\theta_p = \theta_{min}$ and the pth order will be missing. Owing to the regularity of the pattern, if $p = 2$ all the even orders will be missing and if $p = 3$ the 3rd, 6th, etc, orders will be missing.

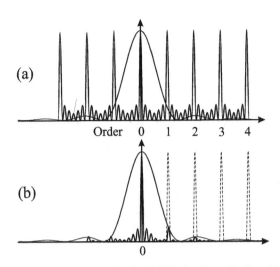

Figure 7.4. (a) Intensity of diffraction orders assuming infinitesimally small slits with a single slit diffraction pattern superimposed. (b) The effect of the single slit diffraction is to reduce the intensity in the higher orders.

7.3 Diffraction grating performance

7.3.1 The diffraction grating equation

Equation (7.4) is the basic diffraction grating equation giving the distribution of intensity, $I(\theta)$, as a function of angle and gives the positions of the principal maxima (diffracted orders), as a result of the dependence of δ on θ. The maxima occur at $\delta = 0$, $n2\pi$, where n, the order of diffraction, is an integer. (Note that n is also the number of wavelengths in the path difference between waves from adjacent slits in the grating at angle θ.) This equation gives us the angular distribution of the diffracted orders for monochromatic light, so we see that, even when the spectrum of the light has an infinitesimal spread in wavelength, the diffracted light will spread out over a finite range of angles. For simplicity, we consider normal incidence on the grating. Then principal maxima occur at angles given by,

$$d \sin \theta = n\lambda. \tag{7.6}$$

In the case where the plane waves are incident on the grating at an angle $\pm\phi$ to the normal, an extra path difference between the light from adjacent slits will be introduced given by $\pm d \sin \phi$, as shown in figure 7.5. The more general diffraction grating equation is then,

$$d(\sin \theta \pm \sin \phi) = n\lambda. \tag{7.7}$$

7.3.2 Angular dispersion

The angular dispersion is an important parameter for a spectrograph because it specifies its ability to separate different spectral components. The angular

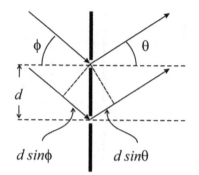

Figure 7.5. Plane wave incident from the left on a grating with slits separated by d at angle ϕ, with diffracted beams at angle θ, showing the path difference between adjacent diffracted beams. The angles shown are positive with respect to the normal.

separation, $d\theta$, between spectral components differing in wavelength by $d\lambda$, is found by differentiating equation (7.6),

$$\frac{d\theta}{d\lambda} = \frac{n}{d \cos \theta}.$$

(7.8)

7.3.3 Resolving power

One of the most important properties of an instrument used to study the spectrum of light is the resolving power, i.e. the ability of the instrument to distinguish unambiguously the difference between two spectral components that have wavelengths that differ by only a small amount. Since the diffraction will produce a finite angular spread, for even a monochromatic light source, it will be necessary to ensure that the instrument will produce an angular separation of two closely spaced, monochromatic spectral components that is greater than the angular spread of each. There is a conventional definition of the condition that allows such differentiation, which demands that the principal maximum arising from one spectral component lies at the position of the first minimum on the side of the principal maximum of the other component as shown in figure 7.6(a).

Principal maxima for wavelength, λ, occur for a phase difference of $\delta = n2\pi$. From equation (7.4) we see that the first minimum after the principal maximum occurs when $N\delta/2 = n2\pi \pm \pi$. The change in the phase difference, δ, between a maximum and the first minimum is therefore, $\Delta\delta_{\min}$,

$$\Delta\delta_{\min} = \pm\frac{2\pi}{N}.$$

(7.9)

The difference in the patterns for different wavelengths will be observed as a difference in the diffraction angle for the two components shown in figure 7.6(b). We express the resolution criterion in terms of the relation between phase difference, δ, and angle, θ, given by equation (7.2),

$$\delta = \frac{2\pi}{\lambda}d \sin \theta.$$

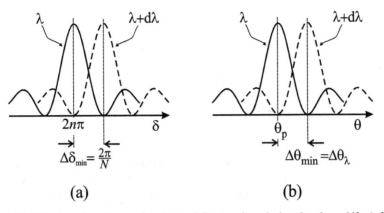

Figure 7.6. (a) Principal maxima for wavelength λ and for $\lambda + d\lambda$ such that the phase shift, δ, for the two wavelength components differs by the change in δ between the maxima and first minima. (b) The same situation plotted as a function of diffraction angle θ. The angular width to the first minimum $\Delta\theta_{\min}$ equals the angular separation $\Delta\theta_\lambda$ between the two wavelengths.

We find the angular spread to the first minimum, $\Delta\theta_{\min}$, by differentiating the phase with respect to angle, θ,

$$\frac{d\delta}{d\theta} = \frac{2\pi}{\lambda}d \cos \theta. \tag{7.10}$$

Hence the small angle, $\Delta\theta$, corresponding to this small phase change, $\Delta\delta$, is given by,

$$\Delta\delta_{\min} = \frac{2\pi}{\lambda}d \cos \theta\Delta\theta_{\min}. \tag{7.11}$$

Then, from equations (7.9) and (7.11),

$$\Delta\delta_{\min} = \frac{2\pi}{\lambda}d \cos \theta \cdot \Delta\theta_{\min} = \frac{2\pi}{N}.$$

Hence, the angular width of the peak to the first minimum is,

$$\Delta\theta_{\min} = \frac{\lambda}{Nd \cos \theta}. \tag{7.12}$$

The angular separation, $\Delta\theta_\lambda$, of principal maxima for λ and $\lambda + \Delta\lambda$ is found from equation (7.8),

$$\frac{d\theta}{d\lambda} = \frac{n}{d \cos \theta}.$$

Hence,

$$\Delta\theta_\lambda = \frac{n}{d \cos \theta}\Delta\lambda. \tag{7.13}$$

The resolution criterion is defined by equating the angular separation of the spectral components, $\Delta\theta_\lambda$, to the angular separation of the peak and the first minimum, $\Delta\theta_{\min}$,

$$\Delta\theta_\lambda = \Delta\theta_{\min}.$$

Hence, from equations (7.12) and (7.13), the resolving power, R.P., is,

$$\frac{\lambda}{\Delta\lambda} = nN. \tag{7.14}$$

7.3.4 Free spectral range

The diffraction grating equation that specifies the angle between the grating normal and each order of diffraction, equation (7.6), shows that two different wavelengths, λ, $\lambda + \Delta\lambda$ may have the same diffraction angle if they are in different orders. The wavelength difference in this case is known as the *free spectral range*, FSR. Specifically, the nth order of λ and the $(n + 1)$th order of $(\lambda - \Delta\lambda_{FSR})$ may lie at the same angle, θ. For light incident normally on the grating ($\phi = 0$), we have,

$$n\lambda = d \sin\theta = (n + 1)(\lambda - \Delta\lambda).$$

Hence, overlap occurs for these wavelengths at this angle and the FSR is therefore

$$\Delta\lambda_{FSR} = \frac{\lambda}{(n + 1)}. \tag{7.15}$$

If the spectral range being examined exceeds the FSR then confusion is created in the recorded spectrum by overlapping orders. This scenario may be avoided by using a grating with a suitably large FSR. However, there is no free lunch, because, recalling that the R.P. is proportional to n (equation (7.14)), and the FSR is proportional to $1/n$, (equation (7.15)), we see that improving the free spectral range comes at the cost of reducing the resolving power.

7.4 Blazed (reflection) gratings

We observed that the effect of diffraction at the slits of finite width can result in a reduction of light intensity at certain angles, as shown in figure 7.4. Usually, most of the light falling on a reflection grating is reflected into the zero order, as shown in figure 7.7(a), leaving less light for the diffracted orders. This reduction in brightness of the diffracted spectrum can be mitigated by the use of 'blazed' reflection gratings. The reflecting surfaces, acting as the slits in a transmission grating, can be set at an angle, the blaze angle, ξ, so that the reflected light is sent in the same direction as the diffracted order of choice for a given wavelength. For an angle of incidence, ϕ, and diffracted angle, θ, the blaze angle shown in figure 7.7(b) will be,

$$\xi = \frac{1}{2}(\phi + \theta)$$

where ϕ and θ satisfy the grating equation,

$$d(\sin\theta \pm \sin\phi) = n\lambda.$$

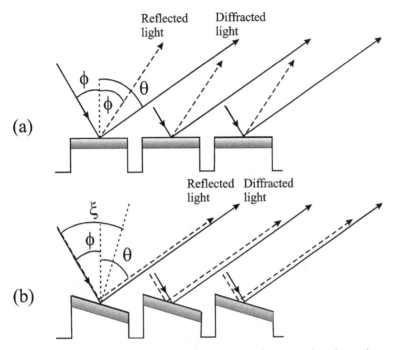

Figure 7.7. (a) Diffraction angle θ differs from the reflection angle ϕ for a normal grating so that most light is reflected at angle ϕ. (b) Blazed grating reflects light at same angle as a diffracted order thus increasing the brightness of this order.

The effect of this blaze is to increase the intensity of the light diffracted into the selected order as shown in figure 7.8 relative to the spectrum formed by an unblazed grating shown in figure 7.4(b).

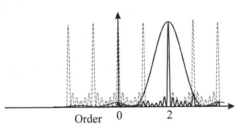

Figure 7.8. Grating intensity pattern for blazed grating with the blaze angle set to reflect light into 2nd order.

7.5 Effect of slit width on resolution and illumination

Consider the image forming system, shown in figure 7.9(a), consisting of two lenses of focal length f_1 and f_2. The image of a slit of width, Δx_s, has a width given by,

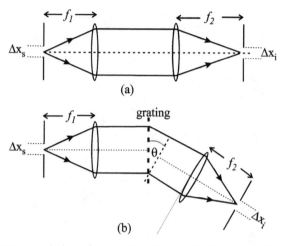

Figure 7.9. (a) Lens system producing an image of a slit of width Δx_s, of width, Δx_i. (b) The spectrally dispersed image at angle θ is foreshortened by $\cos \theta$.

$$\Delta x_i = \left(\frac{f_2}{f_1}\right) \Delta x_s. \tag{7.16}$$

In a diffraction grating spectrograph, shown in figure 7.9(b), using this imaging system, the image of a slit is viewed at the diffraction angle θ, and so the width of the image is foreshortened by $\cos \theta$,

$$\Delta x_i = \left(\frac{f_2}{f_1 \cos \theta}\right) \Delta x_s. \tag{7.17}$$

The minimum resolvable wavelength difference, $\Delta \lambda_R$, has an angular width $\Delta \theta_R$,

$$\Delta \theta_R = \left(\frac{n}{d \cos \theta}\right) \Delta \lambda_R. \tag{7.18}$$

Wavelengths having difference $\Delta \lambda_R$, are separated in the image plane of a lens of focal length, f_2, by Δx_R,

$$\Delta x_R = \frac{f_2 \lambda}{N d \cos \theta}, \tag{7.19}$$

where we used, from equation (7.14), $\Delta \lambda_R = \dfrac{\lambda}{nN}$.

Resolution is achieved, provided $\quad \Delta x_i \leqslant \Delta x_R$.

The limiting slit width, Δx_s, is then,

$$\Delta x_s \leqslant \frac{f_1 \lambda}{N d} \quad \text{or} \quad \Delta x_s \leqslant \frac{f_1 \lambda}{W}. \tag{7.20}$$

Note: the optimum slit width is such that the diffraction pattern of the slit just fills the grating aperture, $W = Nd$. The selection of the appropriate slit width is constrained by the following conditions,

$\Delta x_s > \Delta x_R$: resolution reduced by overlap of images at different wavelengths.

$\Delta x_s < \Delta x_R$: resolution not improved beyond diffraction limit but image brightness is reduced.

The point of this exercise is to compare the size of the slit image in the 'real' spectrograph with the size predicted using the theoretical resolving power. The key difference between the two is that the image in the *real* spectrograph, because it is observed at the diffraction angle θ, is foreshortened by a factor $\cos \theta$. So this means that we have to make the observed image size, Δx_i, larger by a factor $1/\cos \theta$ in order to compare like with like. The theoretical resolving power leads us to a value for the angular separation of wavelengths that differ by $\Delta \lambda_R$. Using our expression for the angular dispersion we find this angular separation to be,

$$\Delta \theta_R = \left(\frac{n}{d \cos \theta} \right) \Delta \lambda_R.$$

This is equation (7.18). We now need to find the spatial extent of the image, Δx_R, corresponding to this theoretical resolution and this is given simply by the relation, $f \Delta \theta$, where the focal length in this case is f_2,

$$\Delta x_R = f_2 \left(\frac{n}{d \cos \theta} \right) \Delta \lambda_R$$

then using $\Delta \lambda_R = \dfrac{\lambda}{nN}$,

$$\Delta x_R = \frac{f_2 \lambda}{Nd \cos \theta}.$$

This is equation (7.19). We note that this calculated 'theoretical' image size does not include any foreshortening effect. Now consider the size of the image in the 'real' spectrograph that *will* exhibit foreshortening. The size of the image formed in the lens imaging system of the spectrograph is determined by the linear magnification of the lens system, $M = v/u = f_2/f_1$.

So the size of any image is related to the entrance slit width, Δx_s, by the relation,

$$\Delta x_i = \frac{f_2}{f_1} \Delta x_s.$$

When an image is viewed in the diffracted beam at angle θ, there is a foreshortening by a factor of $\cos \theta$, i.e. $\Delta x_i' = \Delta x_i \cos \theta$. Conversely any image of size, $\Delta x_i'$, viewed in the diffracted beam corresponds to an unforeshortened image that will be larger by the same factor, i.e. $1/\cos \theta$,

$$\Delta x_i = \frac{f_2 \Delta x_s}{f_1 \cos \theta}.$$

This is equation (7.17). It is this image size—without the effect of foreshortening—that needs to be compared to the theoretical image size. In order to achieve the theoretical resolving power we need to have $\Delta x_i = \Delta x_R$.

In which case,

$$\frac{f_2 \, \Delta x_s}{f_1 \cos \theta} = \frac{f_2 \lambda}{Nd \cos \theta}.$$

Or the limiting slit width is,

$$\Delta x_s \leqslant \frac{f_1 \lambda}{Nd},$$

which is equation (7.20). We note that if the slit is less than this limiting value the resolution is not improved—the image simply becomes less bright—we have diffraction limited imaging.

Optics
The science of light
Paul Ewart

Chapter 8

The Michelson (Fourier transform) interferometer

The Michelson interferometer is a two-beam interference device in which the interfering beams are produced by division of amplitude at a 50:50 beam splitter. Two beams, as we have seen in the case of Young's slits, will produce a cosine-squared intensity pattern. In order to see how this can be used to analyse the spectrum of light, we need to find out how the pattern depends on wavelength, or wavenumber. First of all, we look at the design and operation of the instrument.

8.1 Michelson interferometer

The arrangement used to divide the amplitude into two beams is shown in figure 8.1. The light from an extended source is split by a 50% reflecting and transmitting mirror, i.e. a beam splitter, and travels to a mirror at the end of each arm of the instrument.

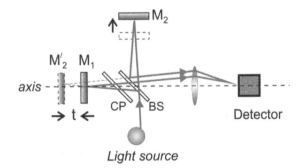

Figure 8.1. The Michelson interferometer. The beam splitter BS sends light to mirrors M_1 and M_2 in arms differing in length by t. M'_2 is the image of M_2 in M_1 giving effectively a pair of parallel reflecting surfaces illuminated by an extended source as in figure 6.6. CP is a compensating plate ensuring that the beams have equal thickness of glass in both arms.

doi:10.1088/2053-2571/ab2231ch8

The distance from the beam splitter to the mirrors, M_1 and M_2, differs by t in the two paths and this corresponds to an optical path difference of $2t = x$. The resulting phase difference between the beams is,

$$\delta = \frac{2\pi}{\lambda}2t \cos \alpha = \frac{2\pi}{\lambda}x \cos \alpha, \tag{8.1}$$

where α is the angle of the interfering beams to the axis. This arrangement is effectively that of light from an extended source reflected in two parallel surfaces, which was discussed in chapter 6, section 6.2.2(b) and illustrated schematically in figure 6.6.

Constructive interference occurs when the phase difference $\delta = 2p\pi$, where p is an integer,

$$x \cos \alpha = p\lambda. \tag{8.2}$$

On axis ($\alpha = 0$) the order of interference is $p = x/\lambda$.

The symmetry of the arrangement gives circular fringes centred on the axis. The fringes are of equal inclination and localised at infinity. They are viewed, therefore, in the focal plane of a lens. The size of the fringes may be deduced by considering the fringe of order, p, which will have a radius, r_p, in the focal plane of a lens with focal length, f (see section 6.2.2(b)).

$$r_p^2 - r_{p+1}^2 = \frac{2f^2\lambda}{x}. \tag{8.3}$$

The two-beam interference pattern is the familiar cosine-squared function,

$$I(x) = I(0)\cos^2\left(\frac{\delta}{2}\right).$$

This may be re-written as,

$$I(x) = \frac{1}{2}I(0)[1 + \cos 2\pi\bar{\nu}x] \tag{8.4}$$

where $\bar{\nu} = \frac{1}{\lambda}$ is the wavenumber.

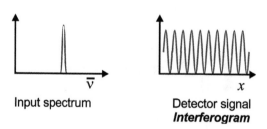

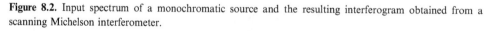

Input spectrum	Detector signal
	Interferogram

Figure 8.2. Input spectrum of a monochromatic source and the resulting interferogram obtained from a scanning Michelson interferometer.

This expression (8.4) describes how the intensity of the interference pattern on the axis of the system varies as the path difference, x, changes. This is usually effected by translating one of the mirrors (M_1 in figure 8.1) along the axis whilst maintaining its orientation perpendicular to the axis. The resulting interferogram, recorded by the detector, is shown schematically in figure 8.2.

8.2 Resolving power of the Michelson spectrometer

If we want to use the Michelson interferometer to resolve two spectral lines that differ in wavelength by $\Delta\lambda$, we need to consider what the resolving power of the instrument needs to be. For wavelengths λ_1 and λ_2, the corresponding wavenumbers are $\bar{\nu}_1$ and $\bar{\nu}_2$. Since these waves originate from different atoms or molecules in the source they have no fixed phase relationship, i.e. they are *incoherent*. The light at each wavelength will produce its own independent interferogram and so we simply add the intensities of the two patterns. The sum is given by,

$$I(x) = \frac{1}{2}I_0(\bar{\nu}_1)[1 + \cos 2\pi\bar{\nu}_1 x] + \frac{1}{2}I_0(\bar{\nu}_2)[1 + \cos 2\pi\bar{\nu}_2 x].$$

For simplicity we will assume, for the moment, that the two components have equal intensity, so $I_0(\bar{\nu}_1) = I_0(\bar{\nu}_2) = I_0(\bar{\nu})$, where $I_0(\bar{\nu})$ is the intensity of each interferogram at $x = 0$ (see figure 8.3).

Then the resulting interferogram is given by,

$$I(x) = I_0(\bar{\nu})\left[1 + \cos 2\pi\left(\frac{\bar{\nu}_1 + \bar{\nu}_2}{2}\right)x \cos 2\pi\left(\frac{\bar{\nu}_1 - \bar{\nu}_2}{2}\right)x\right]. \tag{8.5}$$

This looks like an interferogram of a light source with mean wavenumber, $(\bar{\nu}_1 + \bar{\nu}_2)/2$, multiplied by an envelope function, $\cos 2\pi\ [(\bar{\nu}_1 - \bar{\nu}_2)/2]x$. This envelope

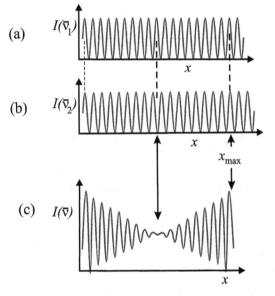

Figure 8.3. (a) Interferogram of source component $\bar{\nu}_1$. (b) Interferogram of source component $\bar{\nu}_2$. (c) Interferogram of combined light showing added intensities (a) and (b). Note, in this case for equal intensity components, the visibility of the fringes cycles to zero and back to unity. To resolve the complete cycle requires a path difference, x_{max}.

function goes first to a zero when a 'peak' of interferogram for $\bar{\nu}_1$ first coincides with a zero in the interferogram for $\bar{\nu}_2$ (or vice versa). The visibility (or contrast) of the fringes cycles to zero and back to unity; this is the tell-tale sign of the presence of the two wavelength components. This cycling is similar to the beating of two sound waves leading to a periodic variation in the intensity of the sound at the difference frequency between the two waves. The number of fringes in the range covering the cycle is determined by the wavenumber difference, $\Delta\bar{\nu} = \bar{\nu}_1 - \bar{\nu}_2$. The instrument will have the power to resolve these two wavenumbers (wavelengths) if the maximum path difference available, x_{max}, is just sufficient to record this cycle in the envelope of the interferogram. The minimum wavenumber difference, $\Delta\bar{\nu}_{min}$, that can be resolved is therefore found from the value of x_{max} characterising the cycle in the cosine envelope function. The envelope function, $\cos 2\pi\,[(\bar{\nu}_1 - \bar{\nu}_2)/2]x$, goes to its first zero when the phase factor is π. The maximum value of the path difference, x_{max}, will determine the minimum value for the wavenumber difference that can be resolved, $\bar{\nu}_1 - \bar{\nu}_2 = \Delta\bar{\nu}_{min}$,

$$2\pi\left(\frac{\Delta\bar{\nu}_{min}}{2}\right)x_{max} = \pi$$

$$\Delta\bar{\nu}_{min} = \frac{1}{x_{max}}. \tag{8.6}$$

This minimum resolvable wavenumber difference determines the resolving power of the instrument and gives an effective measure of the *instrument width*, $\Delta\bar{\nu}_{Inst}$,

$$\Delta\bar{\nu}_{Inst} = \frac{1}{x_{max}}. \tag{8.7}$$

Hence the resolving power, RP, in terms of the minimum resolvable wavelength difference is,

$$\mathrm{RP} = \frac{\lambda}{\Delta\lambda_{Inst}} = \frac{\bar{\nu}}{\Delta\bar{\nu}_{Inst}} = \frac{x_{max}}{\lambda}. \tag{8.8}$$

It is interesting to note that the resolving power of this instrument is clearly related to its physical size. This, incidentally, is not an accident and is a general property of all interferometers, but we will return to this in the next section. When the maximum path difference is made very great, the instrument can be capable of measuring very tiny changes. Instead of measuring a change in wavelength, by using a fixed single-wavelength source, a small change in the length of the interferometer arms can be measured by the shift in the fringe pattern resulting from a length change. Such an instrument, with arms of length 3 or 4 km, and a highly frequency-stable laser source of light, is the basis for the devices used to detect gravitational waves. The Laser Interferometer Gravitational-Wave Observatory (LIGO) instrument in the USA and a similar instrument, Virgo, in Europe registered the differential distortion of space-time in the two arms of the instruments during the passage of gravitational waves from the merger of two black holes. The change in the path difference was about one thousand times smaller than the width of a proton!

8.3 The Fourier transform spectrometer

In figure 8.1 we see that the interferogram looks like the Fourier transform of the intensity spectrum. The interferogram formed by the light of two wavenumbers $\bar{\nu}_1$ and $\bar{\nu}_2$ is,

$$I(x) = \frac{1}{2}I_0(\bar{\nu}_1)[1 + \cos 2\pi\bar{\nu}_1 x] + \frac{1}{2}I_0(\bar{\nu}_2)[1 + \cos 2\pi\bar{\nu}_2 x].$$

In the case of multiple discrete wavelengths,

$$I(x) = \sum_i \frac{1}{2}I_0(\bar{\nu}_i)[1 + \cos 2\pi\bar{\nu}_i x],$$

or,

$$I(x) = \sum_i \frac{1}{2}I_0(\bar{\nu}_i) + \sum_i \frac{1}{2}I_0(\bar{\nu}_i)\cos 2\pi\bar{\nu}_i x. \tag{8.9}$$

The first term on the rhs of equation (8.9) is $1/2\ I_0$, where I_0 is the total intensity at $x = 0$ and the second term is a sum of individual interferograms.

Replacing components with discrete wavenumbers by a continuous spectral distribution,

$$I(x) = \frac{1}{2}I_0 + \int_0^\infty S(\bar{\nu})\cos 2\pi\bar{\nu}x \cdot d\bar{\nu} \tag{8.10}$$

where $S(\bar{\nu})$ is the power spectrum of the source describing the variation of intensity with wavenumber. Now $S(\bar{\nu}) = 0$ for $\bar{\nu} < 0$, and so we can change the limits of the integration and re-write the second term,

$$F(x) = \int_{-\infty}^\infty S(\bar{\nu})\cos 2\pi\bar{\nu}x \cdot d\bar{\nu}. \tag{8.11}$$

We can recognise the function on the rhs as a Fourier transform, i.e. $F(x)$ is the cosine Fourier transform of $S(\bar{\nu})$. Comparing this with equation (8.10), we can write the Fourier transform as,

$$\text{F.T.}\{S(\bar{\nu})\} = I(x) - \frac{1}{2}I_0. \tag{8.12}$$

This equation tells us that if we subtract the constant, background term, $1/2\ I_0$, from the recorded interferogram we obtain a function that is the Fourier transform of the spectral distribution function, i.e. the source spectrum. From the properties of Fourier transforms we can recover the actual spectrum function by a Fourier transformation of the interferogram (minus the constant term),

$$S(\bar{\nu}) \propto \text{F.T.}\left\{I(x) - \frac{1}{2}I_0\right\}. \tag{8.13}$$

So, apart from a constant of proportionality, the Fourier transform of the interferogram yields the intensity or power spectrum of the source. Figure 8.2 shows a simple example.

The Michelson interferometer effectively compares a wavetrain with a delayed replica of itself. The maximum path difference that the device can introduce, x_{max}, is therefore the limit on the length of the wavetrain that can be sampled. The longer the length measured, the lower the uncertainty in the value of the wavenumber obtained from the Fourier transform. Distance, x, and wavenumber, $\bar{\nu}$, are Fourier pairs or conjugate variables (see equation (8.11)). This explains why the limit on the uncertainty of wavenumber (or wavelength) measurement, $\Delta\bar{\nu}_{Inst}$, is just the inverse of x_{max}. In essence, this explains the general rule for all interferometers including diffraction grating instruments that the instrument width is inversely proportional to the maximum path difference that can be produced in the instrument,

$$\Delta\bar{\nu}_{Inst} = \frac{1}{\text{Maximum path difference between interfering beams}}. \tag{8.14}$$

In order to be perfectly monochromatic a light wave must extend to infinity. In other words, a delta function in wavenumber space, i.e. an infinitesimally small spread in wavenumber, requires that the corresponding Fourier variable, distance, must extend to infinity. A wavetrain that is finite will correspondingly have a Fourier transform in the wavenumber that will have a finite width. Thus, since the Michelson instrument has a limiting range of path difference, x_{max}, it can produce an interference pattern from only a finite length of the wavetrain. Therefore, the Fourier transform of a monochromatic wave will not be a delta function but have a finite width. This finite width corresponds to the effective instrument width, $\Delta\bar{\nu}_{Inst}$.

8.4 The Wiener–Khinchin theorem

The discussion of the Michelson interferometer above in terms Fourier transforms gives the opportunity for an interesting digression! The operation of the device involves recording an intensity which is the product of two fields, $E(t)$ and its delayed replica, $E(t + \tau)$, integrated over many cycles. (The delay $\tau = x/c$.) The interferogram as a function of the delay may be written,

$$\gamma(\tau) = \int E(t)E(t + \tau)\mathrm{d}t. \tag{8.15}$$

Taking the integral from $-\infty$ to $+\infty$ we define the autocorrelation function of the field to be,

$$\Gamma(\tau) = \int_{-\infty}^{\infty} E(t)E(t + \tau)\mathrm{d}t. \tag{8.16}$$

The autocorrelation theorem states that for a function $E(t)$ with Fourier transform $F(\omega)$,

$$\text{F.T.}\{\Gamma(\tau)\} = |F(\omega)|^2 = F^*(\omega) \cdot F(\omega). \tag{8.17}$$

We note the similarity between the autocorrelation theorem and the convolution theorem. The physical analogue of the autocorrelation theorem is the *Wiener–Khinchin theorem* (or *Wiener–Kintchine theorem*), which can be stated as follows,

The Fourier transform of the autocorrelation of a signal is the spectral power density of the signal.

Now we have seen that the Michelson interferogram is just the autocorrelation of the light wave (signal). The autocorrelation theorem is stated in terms of frequency but we note that ω and $\bar{\nu}$ are related by a factor, $2\pi c$, where c is the speed of light. The Wiener–Khinchin theorem is a more general result and we have arrived at it by considering the special case of interference in the Michelson or Fourier transform interferometer.

8.5 Fringe visibility

8.5.1 Fringe visibility and relative intensities

Figure 8.3(c) shows an interferogram made up of two independent sources of different wavelengths. The contrast of the fringes, i.e. the difference in intensity between the maximum, I_{max}, and minimum, I_{min}, intensity, varies with increasing path difference, x. We define the 'visibility' of the fringes, V, by,

$$V = \frac{I_{max} - I_{min}}{I_{max} + I_{min}}. \tag{8.18}$$

The fringe visibility 'comes and goes' periodically as the two independent patterns get into and out of step. The example shown consisted of two sources of equal intensity. The visibility varies between 1 and 0. If, however, the two components had different intensity, $I_1(\bar{\nu}_1)$ and $I_2(\bar{\nu}_2)$, then the envelope function of the interferogram does not go to zero. The contrast of the fringes varies from a maximum of $I_1(\bar{\nu}_1) + I_2(\bar{\nu}_2)$ at zero path difference (or time delay) to a minimum value of $I_2(\bar{\nu}_1) - I_2(\bar{\nu}_2)$. Denoting the intensities by I_1 and I_2,

$$V_{max} = \frac{(I_1 + I_2) - 0}{(I_1 + I_2) + 0}; \quad V_{min} = \frac{I_1 - I_2}{I_1 + I_2}.$$

Hence,

$$\frac{V_{min}}{V_{max}} = \frac{I_1 - I_2}{I_1 + I_2},$$

which leads to,

$$\frac{I_1}{I_2} = \frac{1 + V_{min}/V_{max}}{1 - V_{min}/V_{max}}. \tag{8.19}$$

Thus, measuring the ratio of the minimum to maximum fringe visibility, V_{min}/V_{max}, allows the ratio of the two intensities to be determined.

8.5.2 Fringe visibility, coherence and correlation

When the source contains a continuous distribution of wavelengths (or wavenumbers) the visibility decreases to zero with increasing path difference, x, and never recovers.

The two parts of each of the Fourier components (individual frequencies) in each arm of the interferometer are in phase at zero path difference (zero time delay). At large path differences there will be a continuous distribution of interferograms with a range of phase differences that 'average' to zero and no steady state fringes are visible. The path difference, x_o, introduced that brings the visibility to zero is a measure of the wavenumber difference, $\Delta \bar{\nu}_L$, across the width of the spectrum of the source,

$$\Delta \bar{\nu}_L \approx \frac{1}{x_o} \tag{8.20}$$

where $\Delta \bar{\nu}_L$ is the spectral linewidth of the source.

All light sources have a finite spectral linewidth and may be thought of as emitting wavetrains of a finite average length. When these wavetrains are split in the Michelson interferometer, and recombined after a delay, interference will occur only if some parts of the wavetrains overlap. Once the path difference, x, exceeds the average length of wavetrains no further interference is possible. The two parts of the divided wavetrain are no longer 'coherent'. The Michelson interferogram thus gives us a measure of the degree of coherence in the source. A perfectly monochromatic source (if it existed!) would give an infinitely long wavetrain and the visibility would be unity for all values of x. The two parts of the divided wavetrain in this case remain perfectly correlated after any delay is introduced. If the wavetrain has random jumps in phase separated in time on average by say, τ_c, then, when the two parts are recombined after a delay $\tau_d < \tau_c$, only part of the wavetrains will still be correlated. The wavetrains from the source stay correlated with a delayed replica only for the time, τ_c, which is known as the *coherence time*. Thus, we see that the Michelson interferogram provides us with the *autocorrelation function* or self-correlation along the length of the electromagnetic wave emitted by the source (see section 5.3). Thus, the Michelson provides a measure of the *'longitudinal coherence'*.

Light sources may also be characterised by their *'transverse coherence'*. This is a measure of the phase correlation that the waves exhibit in a plane transverse to the direction of propagation. Monochromatic light emanating from a 'point' source will give spherical wave-fronts, i.e. every point on a sphere centred on the source will have the same phase. Similarly, a plane wave is defined as a wave originating effectively from a point source at infinity. Such a source will provide Young's slit interference no matter how large the separation of the slits. (The fringes, of course, will be very tiny for large separations.) If the slits are illuminated by two separate point sources, with the same monochromatic wavelength but with a small displacement, then two sets of independent fringes are produced. The displacement of the sources determines the displacement of the two patterns. For small slit separation this may be insignificant and fringes will be visible. When, however, the slit separation is increased the 'peaks' of one pattern overlap the 'troughs' of the other pattern and uniform illumination results. The separation of the slits in this case thus indicates the extent of the spatial correlation in the phase of the two monochromatic sources, i.e. this measures the *transverse coherence* in the light from the extended source.

Optics
The science of light
Paul Ewart

Chapter 9

The Fabry–Pérot interferometer

This instrument uses multiple beam interference by division of amplitude. Figure 9.1 shows a beam from a point on an extended source incident on two reflecting surfaces separated by a distance, d. Note that this distance is the optical distance, i.e. the product of refractive index, n, and the physical length, d. For convenience we will omit n from the equations that follow but it needs to be included when the space between the reflectors is not a vacuum. Multiple beams are generated by partial reflection at each surface resulting in a set of parallel beams having a relative phase, δ, introduced by the extra path, $2d \cos \theta$, between successive reflections. Since the beams are parallel interference occurs at infinity and, since the phase difference depends on the angle, θ, of the beams relative to the axis (see section 6.2.2(b)), the fringes are of equal inclination and localised at infinity. In practice, they are observed in the focal plane of a lens as a pattern of concentric circular rings.

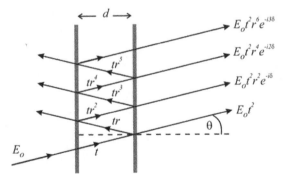

Figure 9.1. Multiple beam interference of beams reflected and transmitted by parallel surfaces. For simplicity in the diagram the coefficient for the amplitude reflection and transmission for the two surfaces are shown as being equal to r and t, respectively.

9.1 The Fabry–Pérot interference pattern

In order to calculate the form of the interference pattern we need to find the sum of all the amplitudes of the interfering beams produced by the multiple reflections at the two parallel surfaces. Figure 9.1 shows a beam of amplitude, E_0, incident on two parallel surfaces at an angle, θ, to the normal, i.e. the axis of the system. Both surfaces in this case are shown as having the same reflection and transmission coefficients, r and t respectively. However, in general, the amplitude reflection and transmission coefficients for the surfaces may be different and we denote them as r_1, t_1, at the first surface (the left side) and r_2, t_2, at the second (right side) surface. The phase difference between successive beams is,

$$\delta = \frac{2\pi}{\lambda} 2d \cos \theta. \tag{9.1}$$

An incident wave, $E_0 e^{-i\omega t}$, is transmitted as a sum of waves with amplitude and phase given by,

$$E_t = E_0 t_1 t_2 e^{-i\omega t} + E_0 t_1 t_2 r_1 r_2 e^{-i(\omega t - \delta)} + E_0 t_1 t_2 r_1^2 r_2^2 e^{-i(\omega t - 2\delta)} + \cdots \text{etc.}$$

Taking the sum of this geometric progression in $r_1 r_2 e^{i\delta}$,

$$E_t = E_0 t_1 t_2 e^{i\omega t} \left[\frac{1}{1 - r_1 r_2 e^{i\delta}} \right]$$

and multiplying by the complex conjugate to find the transmitted intensity,

$$I_t = E_t E_t^* = E_0^2 t_1^2 t_2^2 \left[\frac{1}{1 + r_1^2 r_2^2 - 2r_1 r_2 \cos \delta} \right].$$

Writing $E_0^2 = I_0$, $r_1 r_2 = R$, $t_1 t_2 = T$, and $\cos \delta = (1 - 2 \sin^2 \delta/2)$,

$$I_t = I_0 \frac{T^2}{(1 - R)^2} \left[\frac{1}{1 + \frac{4R}{(1 - R)^2} \sin^2 \delta/2} \right]. \tag{9.2}$$

If there is no absorption in the reflecting surfaces $T = (1 - R)$ then defining the term,

$$\frac{4R}{(1 - R)^2} = \Phi \tag{9.3}$$

we can write,

$$I_t = I_0 \left[\frac{1}{1 + \Phi \sin^2 \delta/2} \right]. \tag{9.4}$$

This is known as the *Airy function*.

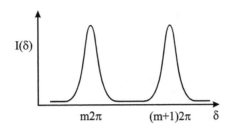

Figure 9.2. The Airy function showing fringes of order m, $m + 1$ as a function of δ.

The Airy function describes how the intensity of the interference fringes changes with the change in the phase factor, δ. The fringe maxima will occur repeatedly when the ($\sin \delta/2$) term in the denominator of equation (9.4) is zero, i.e. when $\delta = m2\pi$, where m is an integer, as shown in figure 9.2,

$$\delta = \frac{2\pi}{\lambda}2d \cos \theta = m2\pi. \tag{9.5}$$

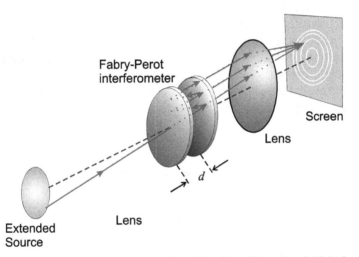

Figure 9.3. Schematic diagram of an arrangement to view Fabry–Pérot fringes. Parallel light from the Fabry–Pérot is focussed on the screen leading to a real image of the fringes that were localised at infinity.

9.2 Observing Fabry–Pérot fringes

According to equation (9.5), the phase factor may be varied by changing d, λ, or θ, and so the fringes of the Airy pattern may be observed as a function of any one of these parameters. A system for viewing many whole fringes is shown in figure 9.3. An extended source of monochromatic light is used with a lens to form the fringes on a screen. Light from any point on the source passes through the F.P. at a range of angles illuminating a number of fringes. The fringe pattern is formed in the focal plane of the lens.

From equation (9.5), the mth fringe is at an angle, θ_m, given by,

$$\cos \theta_m = \frac{m\lambda}{2d}.$$ (9.6)

We usually observe a set of fringes close to the axis where the angles from the axis to the fringes are small. Furthermore, the angular separation of the mth and $(m + 1)$th fringe, $\Delta\theta_m$, will also be small. We can approximate, $\theta_m \approx \theta_{m+1} = \theta$.

For small angles, and using the trigonometric identity for the difference of two cosines, $\cos \theta_{m+1} - \cos \theta_m$, we obtain,

$$\theta\Delta\theta_m = \frac{\lambda}{2d} = \text{constant.}$$

Therefore, as θ gets larger, $\Delta\theta_m$ must get smaller and so the fringes get closer together towards the outside of the pattern. The radius of the fringe at θ_m is,

$$\rho(\lambda) = f\theta_m = f \cos^{-1}\left(\frac{m\lambda}{2d}\right).$$ (9.7)

An alternative method to view fringes is 'centre spot scanning'. A point source at the focal point of a positive lens producing a collimated beam may be used to illuminate the interferometer and the central fringe is imaged on a 'pinhole'. Light transmitted through the pinhole is monitored as a function of d or λ. Fringes are produced of order m, linearly proportional to d or $\bar{v}$, $(1/\lambda)$. This also has the advantage that, because all the available light is collimated, it is all collected in the detected fringe on axis. This can be a particular advantage when dealing with weak sources of light.

9.3 Finesse

The separation of the fringes is 2π in δ-space, and the width of each fringe is defined by the half-intensity point of the Airy function, $I_t/I_0 = 1/2$, i.e. when,

$$\Phi \sin^2 \delta/2 = 1.$$

The value of δ at this half-intensity point is $\delta_{1/2}$,

$$\sin^2\left(\frac{\delta_{1/2}}{2}\right) = \frac{1}{\Phi}.$$

The difference in the phase between the value at a peak and $\delta_{1/2}$ corresponds to a small angle so we have,

$$\delta_{1/2} = \frac{2}{\sqrt{\Phi}}.$$

The full width at half maximum (FWHM) is then $\Delta\delta$,

$$\Delta\delta = \frac{4}{\sqrt{\Phi}}.$$

The sharpness of the fringes may be defined as the ratio of the separation of fringes to the halfwidth, FWHM, and is denoted by the finesse, F,

$$F = \frac{2\pi}{\Delta\delta} = \frac{\pi\sqrt{\Phi}}{2} \tag{9.8}$$

or, using equation (9.3),

$$F = \frac{\pi\sqrt{R}}{(1 - R)}. \tag{9.9}$$

So the sharpness of the fringes is determined by the reflectivity of the mirror surfaces. A large value of finesse will produce sharper fringes and allow higher spectral resolution as we will see below.

Equation (9.9) allows us to calculate the value of the reflectivity coefficient we need to provide the required finesse. This will require solving a quadratic equation for R, but we can get quite a good estimate using the fact that we will usually need a value of R just a little bit less than 1, e.g. $R = 0.9$. We can then approximate equation (9.9) to, $F \sim \frac{3}{(1-R)}$. This approximation is also a useful check that we have found the correct solution of the quadratic equation for R!

9.4 The instrument width

The width of a fringe formed in monochromatic light is the instrument width, $\Delta\delta_{Inst}$. Using the definition of finesse we can write this as,

$$\Delta\delta_{Inst} = \frac{2\pi}{F}. \tag{9.10}$$

A more practically useful expression for the instrument width is, $\Delta\bar{v}_{Inst}$, the instrument width in terms of the apparent spread in wavenumber produced by the instrument for monochromatic light.

For on-axis fringes ($\cos\theta = 1$), we can write,

$$\delta = 2\pi\bar{v}2d$$
$$d\delta = 2\pi2d \ d\bar{v}.$$

Hence,

$$d\delta = \Delta\delta_{Inst} = 2\pi2d \ \Delta\bar{v}_{Inst} = \frac{2\pi}{F} \quad \text{and}$$

$$\Delta\bar{v}_{Inst} = \frac{1}{2dF}. \tag{9.11}$$

As we will see below, the instrument width determines the resolving power and we note that equation (9.11) tells us that this depends again on the physical size of the instrument given by the separation, d, of the reflecting surfaces.

9.5 Free spectral range, FSR

Figure 9.4 shows two successive orders for light having different wavenumbers, $\bar{v}$ and $(\bar{v} - \Delta\bar{v})$. Orders are separated by a change in δ of 2π. The $(m + 1)$th order

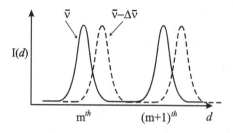

Figure 9.4. Fabry–Pérot fringes for wavenumber $\bar{\nu}$ and $(\bar{\nu} - \Delta\bar{\nu})$ observed in the centre spot scanning mode. The mth-order fringe of $\bar{\nu}$ and of $(\bar{\nu} - \Delta\bar{\nu})$ appear at a slightly different values of the interferometer spacing d. When the wavenumber difference, $\Delta\bar{\nu}$, increases so that the mth-order fringe of $(\bar{\nu} - \Delta\bar{\nu})$ overlaps the $(m + 1)$th order of $\bar{\nu}$, the wavenumber difference equals the free spectral range, FSR.

of wavenumber, $\bar{\nu}$, may overlap the mth-order of $(\bar{\nu} - \Delta\bar{\nu})$, i.e. changing the wavenumber by $\Delta\bar{\nu}$ moves a fringe to the position of the next order of the original wavenumber $\bar{\nu}$,

$$\bar{\nu}2d = (m + 1) \quad \text{and} \quad (\bar{\nu} - \Delta\bar{\nu})2d = m$$
$$\therefore \quad \Delta\bar{\nu}2d = 1.$$

This wavenumber span is called the free spectral range, FSR,

$$\Delta\bar{\nu}_{FSR} = \frac{1}{2d}. \tag{9.12}$$

In figure 9.4 the different orders for each wavelength (wavenumber) are made visible in the centre spot scanning mode, by changing the plate separation d. (Because changing d will change δ.) The phase, δ, can be varied also by changing λ or θ as shown by equation (9.5). In figure 9.3 the different orders for a given wavelength will be distinguished by having different angles, θ, to the axis. If the source emits several different wavelengths, fringes of the same order for each wavelength will have different radii on the screen and so can be distinguished.

9.6 Resolving power

We can now express the instrument width in terms of the FSR and finesse, using equations (9.11) and (9.12),

$$\Delta\bar{\nu}_{Inst} = \frac{\Delta\bar{\nu}_{FSR}}{F} = \frac{1}{2dF}. \tag{9.13}$$

Two monochromatic spectral lines differing in wavenumber by $\Delta\bar{\nu}_R$ are just resolved if their fringes are separated by the instrumental width, $\Delta\bar{\nu}_R = \Delta\bar{\nu}_{Inst}$.

As in figure 9.5, the fringes of the same order for each spectral line separated in wavenumber by $\Delta\bar{\nu}_R$ could be recorded by varying d or θ.

The resolving power is then given by,

$$\frac{\bar{\nu}}{\Delta\bar{\nu}_R} = \frac{\bar{\nu}}{\Delta\bar{\nu}_{Inst}}.$$

Now, $\bar{\nu} = m/2d$,

$$\frac{\bar{\nu}}{\Delta\bar{\nu}_{Inst}} = \frac{m}{2d}\frac{2dF}{1}.$$

Hence the resolving power is,

$$\text{R.P.} = \frac{\bar{\nu}}{\Delta\bar{\nu}_R} = mF. \tag{9.14}$$

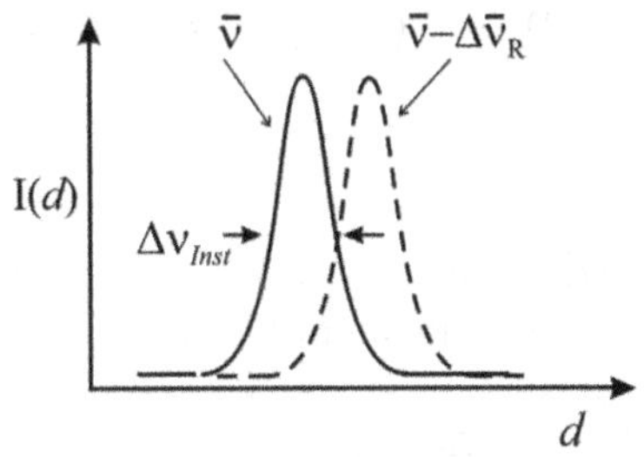

Figure 9.5. Resolution criterion: light of two different wavenumbers $\bar{\nu}$, $\bar{\nu} - \Delta\bar{\nu}_R$ is resolved when the separation of fringes for $\bar{\nu}$ and $\bar{\nu} - \Delta\bar{\nu}_R$ is equal to the instrument width $\Delta\bar{\nu}_{Inst}$.

Note that F defines the effective number of interfering beams and m is the order of interference. This result is therefore similar to the expression for resolving power of a diffraction grating in equation (7.14).

Alternatively, F determines the maximum effective path difference, since it determines the effective number of reflections between the the two surfaces.

$$\text{Maximum path difference} = (2d\cos\theta) \times F \quad \text{and} \quad (2d\cos\theta = m\lambda).$$

So,

$$\frac{\text{Maximum path difference}}{\lambda} = mF. \tag{9.15}$$

In other words, the resolving power, RP, is the number of wavelengths in the maximum path difference. We see again that the resolving power depends on the physical size of the instrument.

9.7 Practical matters

In practice, a Fabry–Pérot interferometer may be constructed with a fixed or variable spacing between two parallel partially reflecting mirrors. A fixed-spacing device, known as an *etalon*, is particularly stable and is made by putting a reflective coating on opposite faces of a parallel sided block of transparent glass or fused silica. We then have to decide what the values of the FSR and the finesse must be for our purpose.

9.7.1 Designing a Fabry–Pérot

(a) FSR: the FSR is usually small and so if a range of different wavelengths are present in the light from the source then it is likely that this range will exceed the FSR. This will produce a complicated fringe pattern and it will be difficult to tell if two fringes are of the same order and different wavelength or the same wavelength and different order. Therefore, Fabry–Pérots are used mostly to determine small wavelength differences between a small number of spectral lines. Suppose a source emits spectral components each of width, $\Delta\bar{\nu}_C$, over a small range, $\Delta\bar{\nu}_S$. We will require $\Delta\bar{\nu}_{FSR} > \Delta\bar{\nu}_S$. This determines the spacing, d, in order to satisfy, $\frac{1}{2d} > \Delta\bar{\nu}_S$ or,

$$d < \frac{1}{2\Delta\bar{\nu}_S}.$$

(b) Finesse: this is determined by the reflectivity of the mirrors. This determines the sharpness of the fringes, i.e. the instrument width. We require,

$$\Delta\bar{\nu}_{Inst} < \Delta\bar{\nu}_c \quad \text{or} \quad \frac{\Delta\bar{\nu}_{FSR}}{F} < \Delta\bar{\nu}_c.$$

Hence,

$$F \geqslant \frac{1}{2d\Delta\bar{\nu}_c}.$$

The required reflectivity R is then found from equation (9.9),

$$F = \frac{\pi\sqrt{R}}{(1 - R)}.$$

9.7.2 Centre spot scanning

Some Fabry–Pérot interferometers are designed to produce a spectrum by varying the separation of the mirrors. This is difficult to do whilst maintaining the plates accurately parallel whilst one of the mirrors is moved to change the separation, d. Fortunately, it is usually necessary to move the mirror by only a small distance as this can easily provide scans over several cycles of the FSR. Such small movements can be effected by using a spacer separating the mirrors made from a piezo-electric material whose thickness can be changed by applying an electrical field. This device produces a set of circular fringes when an extended source is used and it will be necessary to detect the intensity of individual fringes. It is most efficient to measure the intensity of the fringe on the axis, i.e. the centre spot of the pattern, by placing an aperture at this position to transmit only the central fringe. By collimating the light so that most of it travels parallel to the axis, we can ensure most of the available light is used to illuminate only the central fringe. This is particularly advantageous in studying weak light sources. We then need to determine the size of the aperture so that it excludes light from the next order from that on axis. The pinhole admitting the centre spot must be chosen to optimize resolution and light throughput. If it is too large we lose resolution, too small and we waste light and reduce signal-to-noise ratio. We need to calculate the radius of the first fringe away from the central fringe. We recall equation (9.6) that tells us the angular size of the mth-order fringe,

$$\cos \theta_m = \frac{m\lambda}{2d}.$$

If the mth fringe is the central fringe, $\theta_m = 0$, and then $m = 2d/\lambda$. The next fringe has an angular radius,

$$\theta_{m-1} = \cos^{-1}\left(1 - \frac{\lambda}{2d}\right).$$

The fringe radius in the focal plane of lens of focal length, f, will be,

$$\rho_{m-1} = f\theta_{m-1}.$$

This sets the maximum radius of the pinhole to be used.

9.7.3 Limitations on finesse

The sharpness of the fringes is affected if the plates are not perfectly flat. A 'bump' of height, λ/x, is visited by the light reflected between the mirrors multiple times determined by the reflectivity or the finesse. The effective spacing of the mirrors at this point will therefore be different from the nominal value of d, separating the surfaces at other points. The change in the path difference caused by the 'bump' will be $2x$ and so, if the flatness is λ/x, it is not worthwhile making the reflectivity finesse $>x/2$.

We assumed $T = (1 - R)$, i.e. no absorption occurs in the mirror coatings. In practice, however, a small amount of absorption does occur and so,

$$R + T + A = 1,$$

where A is the absorption coefficient of the coatings. The coefficient in equation (9.2) determines the maximum transmitted intensity. When we specifically include the absorption, the coefficient is modified,

$$\frac{T^2}{(1 - R)^2} \Rightarrow \left(\frac{1 - R - A}{1 - R}\right)^2 = \left(1 - \frac{A}{1 - R}\right)^2.$$

Increasing R up to 100% means that $(1 - R) \rightarrow A$ and the coefficient in the Airy function becomes,

$$I_0\frac{T^2}{(1 - R^2)} \Rightarrow 0,$$

i.e. the intensity transmitted to the fringes tends to zero. Therefore it is not practical to increase the reflectivity finesse so much that the fringes become impossible to detect.

9.8 Instrument function and instrument width

The *instrument function* is the 'mathematical' function that describes the shape of the spectrum produced in response to a delta function or monochromatic input. For example, in the case of a diffraction grating spectrometer this is a function made up from a ratio of sine functions, $\sin^2(N\delta/2)/\sin^2(\delta/2)$ given by equation (7.4). This function describes the 'shape' of the interference pattern, i.e. a spectral line.

The *instrument width* is the 'width' of this function defined in some (arbitrary!) way, e.g. in the case of the diffraction grating we choose the separation of the minima on either side of the peak of each order. The pattern produced by a diffraction grating spectrograph consists of lines, sometimes known as *spectral lines*, which are simply images of the entrance slit formed by the light of different wavelengths.

In the case of the Michelson interferometer the instrument does not produce a 'spectral line' directly. Rather the spectral line has to be calculated by finding the Fourier transform of the interferogram. For a monochromatic input wave the interferogram would be a sine wave going on to infinity! The Fourier transform, and hence the instrument function, would then be a delta function. In practice, however, there is a limit to how long the interferogram can be and this is set by the maximum displacement of the mirror. So the interferogram is a sine wave of finite length. The instrument function would then be the Fourier transform of this finite sine wave and will be broader than a delta function, i.e. having some finite width. This function, however, is rarely used. Instead we usually focus on the effective spectral width that the instrument produces for a monochromatic wave—'*the instrument width*'—and this is found from the inverse of the maximum path difference—in units of reciprocal metres.

So, because different instruments produce different shapes of 'spectral lines' from their interference patterns their *instrument functions* are different—so it's hard to compare them. Instead, we usually refer to the *instrument width* of each type of device—Grating, Michelson or Fabry–Pérot, etc, as this allows a more practical way of comparing their usefulness for resolving spectral lines.

Optics
The science of light
Paul Ewart

Chapter 10

Reflection at dielectric surfaces and boundaries

Several of the optical instruments that we have considered so far have used reflecting surfaces or mirrors with varying degrees of reflectivity. The Michelson interferometer used a mirror that reflected 50% of the light intensity and transmitted the other 50%, whereas the mirrors in a Fabry–Pérot might have a reflectivity of over 90% to get high finesse. In some situations we may need to eliminate (as much as possible) any reflection from the surfaces of our optics. To understand how to control the amount of light reflected at a surface, e.g. at an air/glass interface, we begin by reminding ourselves of how electromagnetic waves propagate through dielectric materials and what happens when they cross a boundary between two different media. The wave equations for the electric and magnetic fields in light waves are derived from Maxwell's equations, so that is a good place to begin.

10.1 Electromagnetic waves at dielectric boundaries

The wave equation for the electric and magnetic fields, E and H, are,

$$\nabla^2 E = \varepsilon_o \varepsilon_r \mu_o \mu_r \frac{\partial^2 E}{\partial t^2} \tag{10.1}$$

$$\nabla^2 H = \varepsilon_o \varepsilon_r \mu_o \mu_r \frac{\partial^2 H}{\partial t^2} \tag{10.2}$$

Solutions are of the form,

$$E = E_o \exp i(\mathbf{k} \cdot \mathbf{r} - \omega t)$$

From Maxwell's equations, we also have,

$$\nabla \times E = -\mu_o \mu_r \frac{\partial^2 H}{\partial t^2}$$

doi:10.1088/2053-2571/ab2231ch10

From which, we find,

$$E = {}_2\sqrt{\frac{\mu_o\mu_r}{\varepsilon_o\varepsilon_r}}\,H \quad \text{or} \quad E = \frac{1}{n}{}_2\sqrt{\frac{\mu_o}{\varepsilon_o}}\,H, \tag{10.3}$$

where n is the refractive index of the medium and $\sqrt{\mu_o/\varepsilon_o}$ is the impedance of free space. These equations allow us to relate the amplitudes of the electric and magnetic fields in the wave. Electromagnetic theory also tells us that the amplitudes of each field must satisfy certain boundary conditions that embody the principle of conservation of energy. In particular, the electric and magnetic fields tangential to a surface must be continuous across the boundary. In addition, to account for the possible polarization of the media by the incident fields, the component of the electric displacement field, D, that is perpendicular to the surface, must be continuous across the boundary. The electric displacement field is related to the electric field amplitude by $D = \varepsilon_o\varepsilon_r E$, where ε_o is the permittivity of free space and ε_r is the relative permittivity of the dielectric medium. The relative permittivity of the medium describes how the response of the medium to the electric field of the wave differs from that of the vacuum or free space. The effect of the field is to displace the electronic charges in the atoms or molecules of the medium and so create an array of dipoles that oscillate as they are driven by the electric field of the wave. This results in the wave slowing down as it travels in the material and the change in the speed of the light is described by the refractive index, n_i, of the material. Figure 10.1 shows schematically the incident, reflected and transmitted amplitudes of the electric field of a wave incident normally from a medium of refractive index, n_1, on the surface of medium of index, n_2.

The incident, reflected and transmitted field amplitudes are E_1, E_1' and E_2, respectively. The boundary conditions at the interface of the two media demand that the tangential components of E and H are continuous and this gives us two equations relating the field amplitudes on each side of the boundary,

$$E_1 + E_1' = E_2 \tag{10.4}$$

$$H_1 - H_1' = H_2 \tag{10.5}$$

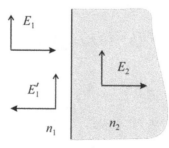

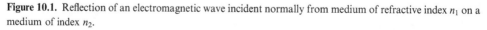

Figure 10.1. Reflection of an electromagnetic wave incident normally from medium of refractive index n_1 on a medium of index n_2.

(Note the minus sign on H' is because the reflected wave is travelling in the opposite direction and so, if the E vector is in the same direction as that of the incident wave, then the magnetic vector must be in the opposite direction—hence the minus sign.) Using equation (10.3) we can rewrite equation (10.5) as,

$$n_1 E_1 - n_1 E_1' = n_2 E_2. \tag{10.6}$$

To find the amplitude reflection coefficient, r, we eliminate the E_2 term using the two simultaneous equations (10.4) and (10.6), and we find (ignoring the negative sign that has no physical significance here),

$$\frac{E_1}{E_1'} = r = \frac{n_2 - n_1}{n_2 + n_1}. \tag{10.7}$$

The intensity reflection coefficient is therefore,

$$R = \left(\frac{n_2 - n_1}{n_2 + n_1}\right)^2. \tag{10.8}$$

For example, for an air/glass interface $R \sim 4\%$.

10.2 Reflection properties of a single dielectric layer

Consider a wave incident from air, refractive index, n_o, on a dielectric layer of index, n_1, deposited on a substrate of refractive index, n_T, as shown in figure 10.2. The incident electric and magnetic wave amplitudes in the air are E_o and H_o, respectively, and E_o', H_o' are the reflected amplitudes. E_1, H_1 and E_1', H_1' are incident and reflected amplitudes in the dielectric layer and E_T is the electric field amplitude transmitted to the substrate.

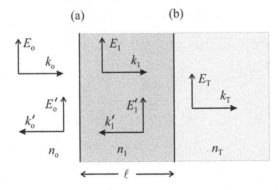

Figure 10.2. Reflected and transmitted waves for a wave incident normally from medium of index n_o on a dielectric layer of thickness ℓ and index n_1 deposited on a substrate of index n_T. (a), (b) The boundaries between the media $n_o{:}n_1$ and $n_1{:}n_T$.

At boundary (a), using the same boundary conditions as before,

$$E_o + E_o' = E_1 + E_1' \tag{10.9}$$

$$H_o - H_o' = H_1 - H_1'. \tag{10.10}$$

Using $H = n_2 \sqrt{\frac{\varepsilon_o}{\mu_o}} E$,

$$n_o(E_o - E_o') = n_1(E_1 - E_1'). \tag{10.11}$$

At boundary (b), E_1 has acquired a phase shift owing to propagating the thickness, ℓ, of the layer,

$$E_1 e^{ik_1\ell} + E_1' e^{-ik_1\ell} = E_T \tag{10.12}$$

and the continuity of the magnetic fields gives,

$$n_1\left(E_1 e^{ik_1\ell} - E_1' e^{-ik_1\ell}\right) = n_T E_T. \tag{10.13}$$

Eliminating E_1 and E_1' from equations (10.9), (10.11), (10.12) and (10.13) we find,

$$E_o + E_o' = \left[\cos k_1\ell - i\left(\frac{n_T}{n_1}\right)\sin k_1\ell\right] E_T \tag{10.14}$$

$$n_o(E_o - E_o') = [-in_1 \sin k_1\ell + n_T \cos k_1\ell] E_T. \tag{10.15}$$

Writing,

$$A = \cos k_1\ell \qquad B = -i\left(\frac{1}{n_1}\right)\sin k_1\ell$$
$$C = -in_1 \sin k_1\ell \qquad D = \cos k_1\ell \tag{10.16}$$

we find, by eliminating E_T,

$$\frac{E_o'}{E_o} = r = \frac{An_o + Bn_on_T - C - Dn_T}{An_o + Bn_on_T + C + Dn_T} \tag{10.17}$$

and, by eliminating E_o',

$$\frac{E_T}{E_o} = t = \frac{2n_o}{An_o + Bn_on_T + C + Dn_T}. \tag{10.18}$$

Now consider the case when $\ell = \lambda/4$, a quarter-wave layer; $k_1\ell = \pi/2$,

$$R = |r|^2 = \left|\frac{n_on_T - n_1^2}{n_on_T + n_1^2}\right|^2. \tag{10.19}$$

For $\ell = \lambda/2$ a half-wave layer; $k_1\ell = \pi$,

$$R = |r|^2 = \left| \frac{n_o - n_T}{n_o + n_T} \right|^2. \tag{10.20}$$

Note that for a half-wave layer the refractive index of the layer does not appear in the reflectivity and the result is the same as for an uncoated surface. (This effect is similar to that of a reflection in a half-wave section of a transmission line.)

10.3 Anti-reflection coatings

An anti-reflection (AR) coating can be made, i.e. one that minimizes the reflection by selecting a dielectric material such that $n_o n_T - n_1^2 = 0$ from equation (10.19). This requires $n_1 = \sqrt[2]{n_o n_T}$. For an air/glass boundary this is not possible; the closest we can get is to have n_1 as low as possible, e.g. MgF$_2$ has $n_1 = 1.38$ giving $R \sim 1\%$.

Physically, we can think of the AR coating working by adding a second reflected wave from the surface of the coating. This second wave is set up so that it is exactly out-of-phase with the reflection from the original glass surface and so cancels out, at least partially, the reflected amplitude. We can improve the performance of an AR coating by adding further reflected waves to give extra cancellation effect. This requires us to use multiple layers and these can be very effective in significantly reducing the reflections from glass surfaces as shown in figure 10.3.

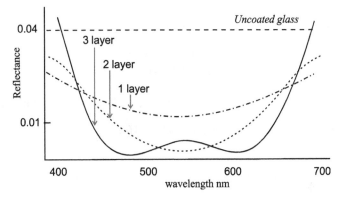

Figure 10.3. Anti-reflection dielectric coatings. A single $\lambda/4$ layer can reduce reflection from 4% to $\sim$1%. Further reduction at specific wavelength regions is achieved by additional layers at the expense of increased reflectivity elsewhere.

Reducing the reflection from the surface of lenses is important to improve the quality of photographic images. Camera lenses are often compound, i.e. constructed using several lenses designed to minimize aberrations in the image. Reflections from multiple surfaces will produce unfocussed light at the image plane, which reduces the contrast in the image. It is possible to reduce the reflection over the important range of the visible spectrum by an appropriate set of layers. However, as shown in figure 10.3, the reduction in reflectivity at certain wavelengths (or colours) is compensated by enhanced reflectivity at other wavelengths. This enhanced reflection at the blue and red end of the spectrum is responsible for the purple hue or blooming on camera or spectacle lenses.

Coatings may also be made to enhance the reflectivity to make high reflectance mirrors. As with the AR coatings, this process is enhanced by using multiple layers, so it is useful to derive a more general way to calculate the reflectivity of a stack of multiple dielectric layers. We use a matrix method as shown in the following section.

10.4 Multiple dielectric layers: matrix method

Write equations (10.14) and (10.15) in terms of r, i.e. E_o'/E_o and t, i.e. E_T/E_o,

$$1 + r = (A + Bn_T)t$$
$$n_o(1 - r) = (C + Dn_T)t$$

or in matrix form,

$$\begin{pmatrix} 1 \\ n_o \end{pmatrix} + \begin{pmatrix} 1 \\ -n_o \end{pmatrix} r = \begin{pmatrix} A & B \\ C & D \end{pmatrix} \begin{pmatrix} 1 \\ n_T \end{pmatrix} t. \tag{10.21}$$

The coefficients A, B, C and D are defined in equation (10.16) and we define the characteristic matrix as,

$$M = \begin{pmatrix} A & B \\ C & D \end{pmatrix}. \tag{10.22}$$

The characteristic matrix for a $\lambda/4$ layer of index, n_m, is,

$$M_m = \begin{pmatrix} 0 & -i/n_m \\ -in_m & 0 \end{pmatrix}. \tag{10.23}$$

A stack of N layers, such as that shown in figure 10.4, has a characteristic matrix,

$$M_{Stack} = M_1 M_2 M_3 \dots M_N. \tag{10.24}$$

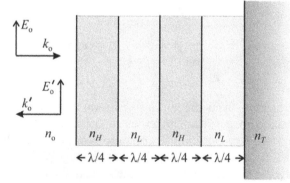

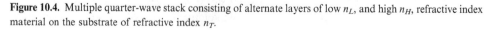

Figure 10.4. Multiple quarter-wave stack consisting of alternate layers of low n_L, and high n_H, refractive index material on the substrate of refractive index n_T.

10.5 High reflectance mirrors

Using equations (10.23) and (10.24), a stack of two dielectric layers of alternate high and low index n_H, n_L, respectively, has the matrix,

$$M_{HL} = \begin{pmatrix} -n_L/n_H & 0 \\ 0 & -n_H/n_L \end{pmatrix}. \tag{10.25}$$

For N such pairs the matrix is M_{HL}^N. From this 2×2 matrix we find the values of A, B, C and D. Hence from equation (10.17) we find the reflectivity of the composite stack,

$$R_{Stack} = \left\{ \frac{1 - \frac{n_T}{n_o}\left(\frac{n_H}{n_L}\right)^{2N}}{1 + \frac{n_T}{n_o}\left(\frac{n_H}{n_L}\right)^{2N}} \right\}^2. \tag{10.26}$$

10.6 Interference filters

A Fabry–Pérot etalon structure may be constructed from two high reflectance stacks separated by a layer that is $\lambda/2$ or an integer multiple of $\lambda/2$, as shown in figure 10.5. The half-wave layer acts as a spacer to determine the FSR. The FSR will therefore be very large such that only one transmission peak may lie in the visible region of the spectrum. This is an interference filter. Filters that transmit a narrower range of the spectrum may be made by increasing the spacer thickness and increasing the reflectance. Extra peaks arising from other orders or wavelengths, may be eliminated using a broad band, high or low pass filter.

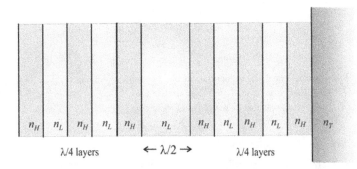

Figure 10.5. Interference filter constructed using multiple dielectric layers consisting of two high-reflectance stacks separated by a $\lambda/2$ layer that acts as a spacer in the Fabry–Pérot type interference device. The spacer may be made in integer multiples of $\lambda/2$ to achieve different values of the FSR.

10.7 Reflection and transmission at oblique incidence

So far we have considered cases where the light waves have been incident on surfaces or boundaries at normal incidence. We now consider the more general case of a light wave striking a dielectric boundary at an angle of incidence, θ, that is not 90°. As with the usual convention, we define the plane of incidence as that plane containing the wave propagation vector, $\underline{k}$, and the normal to the dielectric boundary. This is illustrated in figure 10.6. The incident wave has the form, $E_0 e^{i\omega(t-nr/c)}$, where r is the distance along the propagation axis, OP. In general, the electric field vector, $\underline{E}$, can lie at any angle around the wave vector, $\underline{k}$. However, we can always resolve this vector into two components; one lying in (i.e. parallel to) the plane of incidence, E_P, and one perpendicular to this plane, E_S. When a light wave has its E-vector parallel to the plane of incidence it is referred to as p-polarized light. When the E-vector is perpendicular to this plane it is s-polarized, from the German *senkrecht*, meaning perpendicular. The $\underline{H}$-vectors will be perpendicular to the respective $\underline{E}$-vectors in each case.

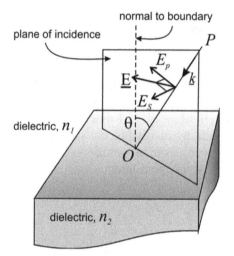

Figure 10.6. Wave with electric field, E, incident at an oblique angle of incidence, θ, from a dielectric with refractive index, n_1, to a dielectric with index, n_2. The electric field, $\underline{E}$ has a component, E_P, parallel to the plane of incidence and, E_S, perpendicular to the plane of incidence.

10.7.1 Reflection and transmission of p-polarized light

We consider first the case of p-polarized light. This is represented in figure 10.7 in which the plane of incidence (the xz-plane) is the plane of the paper. The incident, reflected and transmitted E-fields all lie in this plane and are denoted E_1^P, $E_1^{\vec{P}}$ and E_2^P, respectively, lying at angles θ_1, θ_2 and ϕ, respectively, to the normal or z-axis.

The boundary conditions require,

$$E_1^P \cos\theta_1 e^{i\omega(t-n_1 x \sin\theta_1/c)} - E_1^{\vec{P}} \cos\theta_2 e^{i\omega(t-n_1 x \sin\theta_2/c)} = E_2^P \cos\phi \, e^{i\omega(t-n_2 x \sin\phi/c)} \quad (10.27)$$

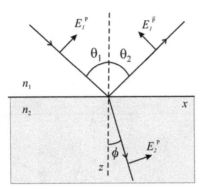

Figure 10.7. Electric field components of p-polarized wave incident at angle θ_1 from a medium of index n_1 to a medium of index n_2. See text for details.

and the incident, reflected and transmitted H-field components out of the paper are,

$$H_1^P + H_1^{\bar{P}} = H_2^P. \tag{10.28}$$

Since these conditions are obeyed at all times the exponential terms must be identical. Hence,

$$\begin{aligned} \theta_1 &= \theta_2 = \theta \\ n_1 \sin\theta &= n_2 \sin\phi. \end{aligned} \tag{10.29}$$

These equations express Snell's law for reflection and refraction.

From section 10.1, we can write $H = nE$ and the boundary conditions then become,

$$\left(E_1^P - E_1^{\bar{P}}\right)\cos\theta = E_2^P \cos\phi$$

$$n_1\left(E_1^P + E_1^{\bar{P}}\right) = n_2 E_2^P.$$

From this we find, using Snell's law (10.29), the ratios of the transmitted and reflected amplitudes to the incident amplitude,

$$\frac{E_2^P}{E_1^P} = \frac{2\cos\theta \sin\phi}{\sin\theta \cos\theta + \sin\phi \cos\phi} = \frac{2\cos\theta \sin\phi}{\sin(\theta+\phi)\cos(\theta-\phi)} \tag{10.30}$$

$$\frac{E_1^{\bar{P}}}{E_1^P} = \frac{\sin\theta \cos\theta - \sin\phi \cos\phi}{\sin\theta \cos\theta + \sin\phi \cos\phi} = \frac{\tan(\theta-\phi)}{\tan(\theta+\phi)}. \tag{10.31}$$

10.7.2 Reflection and transmission of s-polarized light

When the light is polarized such that the E-vector is perpendicular to the plane of incidence, i.e. s-polarized light, the H-fields will lie in this plane as shown in figure 10.8.

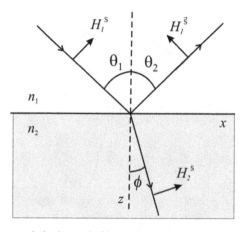

Figure 10.8. Reflection of an s-polarized wave incident at an angle, θ, at a dielectric boundary. The E-vectors are out of the plane of the figure.

Following the same procedure, using the boundary conditions on E and H, we find the following relationships between the incident, reflected and transmitted amplitudes,

$$\frac{E_2^S}{E_1^S} = \frac{2\cos\theta\sin\phi}{\cos\theta\sin\phi + \sin\theta\cos\phi} = \frac{2\cos\theta\sin\phi}{\sin(\theta + \phi)} \qquad (10.32)$$

$$\frac{E_1^{\vec{s}}}{E_1^S} = \frac{\cos\theta\sin\phi - \sin\theta\cos\phi}{\cos\theta\sin\phi + \sin\theta\cos\phi} = -\frac{\sin(\theta - \phi)}{\sin(\theta + \phi)}. \qquad (10.33)$$

Equations (10.30)–(10.33) are the Fresnel equations. They allow us to predict the reflection and transmission coefficients for light of various polarizations incident at any angle on a dielectric surface, i.e. a boundary between two different dielectric media such as air and glass.

The Fresnel equations show that the reflection coefficients for both s- and p-polarized light vary with angle of incidence as shown in figure 10.9.

10.8 Deductions from Fresnel's equations

10.8.1 Brewster's angle

The amplitude reflection coefficient, r, is given for p-polarized light by equation (10.31),

$$\frac{E_1^{\vec{P}}}{E_1^P} = r = \frac{\tan(\theta - \phi)}{\tan(\theta + \phi)}. \qquad (10.34)$$

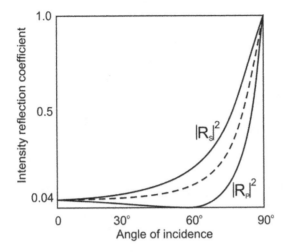

Figure 10.9. Intensity reflection coefficients for s-polarized and p-polarized light as a function of incidence angle from air to glass. The dashed line is the average and represents the behaviour for unpolarized light.

We can see that as $(\theta + \phi)$ approaches $\pi/2$, r will tend to zero as $\tan(\theta + \phi) \to \infty$. In the limit $(\theta + \phi) = \pi/2$, then $\cos \theta = \sin \phi$. The angle of incidence at which this condition is satisfied is known as Brewster's angle, θ_B, and, from equation (10.29),

$$\sin \theta_B = \frac{n_2}{n_1} \sin \phi = \frac{n_2}{n_1} \cos \theta_B$$

or,

$$\tan \theta_B = \frac{n_2}{n_1}. \qquad (10.35)$$

Thus, for p-polarized light incident at angle, θ_B, there will be no reflected wave.

Another consequence of this is that unpolarized light, that consists, as we will see later, of waves with E-vectors varying randomly in all possible orientations, reflected at this angle will become predominantly plane polarized with its E-vector perpendicular to the plane of incidence. This is because any E-vector may be composed of s- and p-polarizations and only the s-polarization will be reflected. An important application of Brewster's angle is in enabling p-polarized light to suffer no reflection losses when passing through a glass window. This is very useful for minimizing reflection losses at window surfaces for intense laser light.

10.8.2 Phase changes on reflection

We first consider the predictions of the Fresnel equations for normal incidence. For $\theta = 0$, using equation (10.30) and equation (10.29) we find,

$$\frac{E_2^P}{E_1^P} = \frac{2\cos\theta\sin\phi}{\sin\theta\cos\theta + \sin\phi\cos\phi}$$

$$= \frac{2\cos\theta(n_1/n_2)\sin\theta}{\sin\theta\cos\theta + \sin\phi\cos\phi} \ .$$

$$= \frac{2(n_1/n_2)}{1 + \frac{\sin\phi\cos\phi}{\sin\theta\cos\theta}} = \frac{2(n_1/n_2)}{1 + \frac{n_1}{n_2}\frac{\cos\phi}{\cos\theta}}$$

Hence,

$$\frac{E_2^P}{E_1^P} = \frac{2n_1}{n_2 + n_1}. \tag{10.36}$$

Similarly, we find,

$$\frac{E_1^{\vec{P}}}{E_1^P} = \frac{n_2 - n_1}{n_2 + n_1} \tag{10.37}$$

$$\frac{E_2^S}{E_1^S} = \frac{2n_1}{n_2 + n_1} \tag{10.38}$$

$$\frac{E_1^{\vec{S}}}{E_1^S} = -\frac{n_2 - n_1}{n_2 + n_1}. \tag{10.39}$$

There appears to be a discrepancy here between the equations (10.37) and (10.39) for the reflected light polarized in two orthogonal planes. At normal incidence, the two planes are indistinguishable, i.e. there is no difference between E_1^P and E_1^S, yet the ratio of the reflected to incident amplitudes have opposite signs! The sign discrepancy arises because we have taken the incident and reflected fields, E_1^P and $E_1^{\vec{P}}$, to be in the opposite directions (see figure 10.7). Assuming $n_2 > n_1$ and, since the ratio $E_1^P/E_1^{\vec{P}}$ is positive, our theory predicts that there is a phase shift of π in the reflected E-vector relative to the incident wave. Note that as the incident and reflected H-vectors are in the same direction there is no phase shift of the H-field. In the case of the s-polarized E-vectors in figure 10.8, they are both perpendicular to the plane of the paper but in opposite directions, i.e. corresponding to a π-phase shift. In the case $n_2 < n_1$ there is no phase shift in E but a π-phase shift occurs in the H-vector. In both cases there is no phase shift in the transmitted wave.

10.8.3 Total (internal) reflection and evanescent waves

When a wave is incident from a medium of index, n_2, obliquely on a less dense medium, index n_1, as shown in figure 10.10, we know that if the angle of incidence exceeds a critical value, θ_{crit}, given by $\theta_{crit} = \sin^{-1}(n_1/n_2)$, then the wave is totally

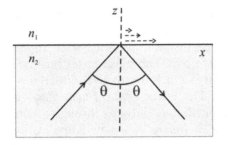

Figure 10.10. Internal reflection, the evanescent wave is shown as dotted lines.

reflected. In this case the angle ϕ, in the less dense medium is an imaginary quantity, leading to phase shifts that lie between 0 and π, and are different for the electric field E, and its corresponding magnetic field H. The analysis is tedious but leads to the following predictions.

1. The amplitude of the reflected beam and incident beams are equal, i.e. total reflection.
2. There is a transmitted beam with the following characteristics,
   (a) it travels parallel to the interface,
   (b) its amplitude decays exponentially with distance perpendicular to the surface,
   (c) it is neither a plane nor a transverse wave—it has a component along the surface,
   (d) its Poynting vector is zero.

The slightly mysterious transmitted wave is known as an *evanescent wave*. Its presence can be detected by bringing another dielectric, say of index also n_2, close to the surface as shown in figure 10.11.

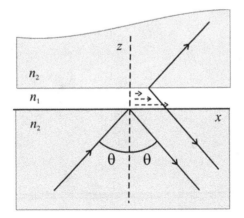

Figure 10.11. Frustrated internal reflection.

When a wave is incident from a medium of index n_2, obliquely on a less dense medium, index n_1, as the gap between the two dielectrics of index n_2 gets smaller (of the order of a few wavelengths), a transmitted wave appears simultaneously with a weakening of the reflected wave. This effect arises because there is a reflection from the second n_1/n_2 boundary, which destructively interferes with the reflected wave from the first boundary. This effect—*optical tunnelling*—is the wave equivalent of quantum mechanical tunnelling of particles through a potential barrier. The effect, known as *frustrated total internal reflection*, finds application in a type of microscopy in which the surface layer of biological samples are excited by evanescent waves. Since these penetrate only a short distance, of the order of 100 nm, fluorescence is induced only in those molecules lying in this thin layer. The fluorescence image is then no longer swamped by the fluorescence from the bulk of the sample that would otherwise overwhelm the signal from the surface layer.

IOP Concise Physics

Optics
The science of light
Paul Ewart

Chapter 11

Polarized light

We have, so far, mostly ignored the direction, or orientation, of the electric field in the light wave and treated the amplitude as a scalar quantity. However, we saw in the last chapter that the way light is reflected at a dielectric boundary is affected by the direction of the electric field relative to the plane of incidence and so we turn now to consider the vector nature of the field. We considered two situations where the light wave had its electric field parallel or perpendicular to the plane of incidence and referred to these as p-polarized and s-polarized, respectively. The polarization of light refers to the direction of the electric field vector, $\mathbf{E}$, of the wave. There are three options for $\mathbf{E}$ for polarized light;

(1) its direction is fixed in space and its amplitude remains constant—*linear* polarization;

(2) its direction rotates at angular frequency, ω, about the direction of propagation and its amplitude remains constant—*circular* polarization;

(3) its direction rotates at angular frequency ω and its amplitude varies between a maximum and minimum during each complete rotation—*elliptical* polarization.

For propagation in the x-direction the vector $\mathbf{E}$ may be resolved into two orthogonal components E_y and E_z. Each of the three polarization states is characterised by a fixed phase relationship between these components. We will consider the states of polarization defined by the nature of this phase relationship. If the phase is randomly varying the light is said to be *unpolarized*.

11.1 Polarization states

An electromagnetic wave travelling in the positive x-direction has an electric field, $\mathbf{E}$, lying at an angle, α, with respect to the y-axis and has components E_y and E_z (see figure 11.1),

doi:10.1088/2053-2571/ab2231ch11

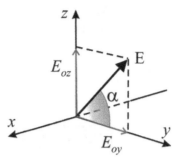

Figure 11.1. Electric field vector in a light wave has components E_{oz} and E_{oy} in a plane orthogonal to the propagation direction along the x-axis.

$$E_y = E_{oy} \cos(kx - \omega t)\underline{\mathbf{j}}$$
$$E_z = E_{oz} \cos(kx - \omega t + \delta)\underline{\mathbf{k}}$$ (11.1)

where δ is the relative phase. The light is *polarized* when δ is a constant.

11.1.1 Case 1: linearly polarized light, $\delta = 0$

The components are in phase. The resultant is a vector, $\mathbf{E}_P$,

$$\mathbf{E}_P = \{E_{oy}\underline{\mathbf{j}} + E_{oz}\underline{\mathbf{k}}\} \cos(kx - \omega t)$$ (11.2)

at a fixed angle, α, to the y-axis,

$$\tan \alpha = \frac{E_{oz}}{E_{oy}}.$$ (11.3)

11.1.2 Case 2: circularly polarized light, $\delta = \pm \pi/2$

Consider, $\delta = -\pi/2$ and $E_{oy} = E_{oz} = E_o$, the components are,

$$E_y = E_o \cos(kx - \omega t)\underline{\mathbf{j}}$$
$$E_z = E_o \sin(kx - \omega t)\underline{\mathbf{k}}$$ (11.4)

and the angle, α, is no longer fixed but given by,

$$\tan \alpha = \frac{\sin(kx - \omega t)}{\cos(kx - \omega t)} = \tan(kx - \omega t).$$ (11.5)

Equation (11.5) tells us that the tip of the **E**-vector rotates at angular frequency, ω, at any position, x, on the axis, and rotates by 2π for every distance, λ, along the x-axis, as shown in figure 11.2. To determine the direction of rotation we can calculate the angle at various positions on the x-axis as follows. Consider the components at position $x = x_o$ and time $t = 0$ to be,

$$E_y = E_o \cos(kx_o)$$
$$E_z = E_o \sin(kx_o).$$

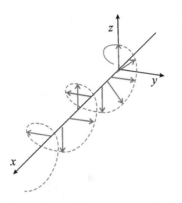

Figure 11.2. Right circularly polarized light propagating in the positive x-direction.

As shown in figure 11.3(a), at this position the vector is at some angle to the y-axis. At this position and at some later time, i.e. at position $x = x_o$ and time $t = kx_o/\omega$, the vector will have rotated to be lying along the y-axis, as shown in figure 11.3(b). So,

$$E_y = E_o$$
$$E_z = 0.$$

When viewed back towards the source the **E**-vector has rotated clockwise (see figure 11.3(b)). This is right circularly polarized light ($\delta = -\pi/2$). *Right* circularly polarized light advances like a *left*-handed screw!

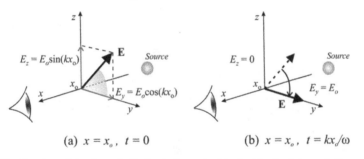

$$\text{(a) } x = x_o, \ t = 0 \qquad\qquad \text{(b) } x = x_o, \ t = kx_o/\omega$$

Figure 11.3. The direction of circular polarization is determined by looking back towards the source. (a) The E-vector at a point $x = x_o$ at time $t = 0$, and (b) the vector at a later time, $t = kx_o/\omega$. In this case the E-vector has rotated clockwise and is denoted right circularly polarized.

Conversely, $\delta = +\pi/2$ is left-circularly polarized light; viewed towards the source the **E**-vector rotates anti-clockwise. Thus, the **E**-vector for right and left circular polarization is written,

$$\mathbf{E}_R = E_o[\cos(kx - \omega t)\underline{\mathbf{j}} + \sin(kx - \omega t)\underline{\mathbf{k}}]$$
$$\mathbf{E}_L = E_o[\cos(kx - \omega t)\underline{\mathbf{j}} - \sin(kx - \omega t)\underline{\mathbf{k}}].$$

$$(11.6)$$

Note that a linear superposition of $\mathbf{E}_R$ and $\mathbf{E}_L$ gives linear or plane polarized light,

$$\mathbf{E}_P = \mathbf{E}_R + \mathbf{E}_L = 2E_o \cos(kx - \omega t)\underline{\mathbf{j}}. \tag{11.7}$$

This is shown schematically in figure 11.4 where the two oppositely rotating components add to give a resultant, $\mathbf{E}_P$, that oscillates between zero and a maximum during the cycle.

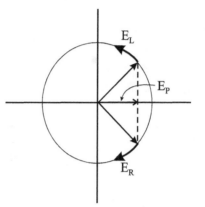

Figure 11.4. Plane polarized light is a superposition of a right- (E_R) and a left- (E_L) circularly polarized component. The resultant vector, $\mathbf{E}_P$, lies, in this case, along the horizontal y-axis.

If the components are of unequal amplitude then the resultant traces out an ellipse, i.e. the light is elliptically polarized, as shown in figure 11.5.

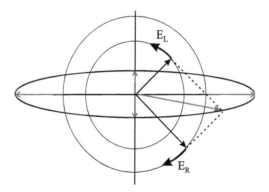

Figure 11.5. A superposition of right- and left-circularly polarized components of unequal magnitude gives elliptically polarized light.

11.1.3 Case 3: elliptically polarized light

In general, there is a relative phase, δ, between y-and z-components which may not be exactly $\pm\pi/2$. From equation (11.1),

$$\begin{aligned}
E_y &= E_{oy} \cos(kx - \omega t) \\
E_z &= E_{oz} \cos(kx - \omega t + \delta).
\end{aligned} \tag{11.8}$$

We may re-write the equation for the E_z component using a trigonometric identity,

$$E_z = E_{oz}[\cos(kx - \omega t)\cos\delta - \sin(kx - \omega t)\sin\delta]. \tag{11.9}$$

Substituting in equation (11.9) using,

$$\cos(kx - \omega t) = \frac{E_y}{E_{oy}}, \quad \text{and} \quad \sin(kx - \omega t) = \left[1 - \left(\frac{E_y}{E_{oy}}\right)^2\right]^{1/2}$$

we obtain,

$$\frac{E_y^2}{E_{oy}^2} + \frac{E_z^2}{E_{oz}^2} - 2\frac{E_y}{E_{oy}}\frac{E_z}{E_{oz}}\cos\delta = \sin^2\delta. \tag{11.10}$$

This is a general equation describing an ellipse and, for $\delta = \pm\pi/2$,

$$\frac{E_y^2}{E_{oy}^2} + \frac{E_z^2}{E_{oz}^2} = 1. \tag{11.11}$$

This is the equation for an ellipse with E_{oy}, E_{oz} as the major/minor axes, disposed symmetrically about the y/z-axes.

For $\delta \neq \pm\pi/2$, the axes of symmetry of the ellipse are rotated relative to the y/z-axes by an angle, θ, given by,

$$\tan 2\theta = 2\frac{E_{oy}E_{oz}}{E_{oy}^2 - E_{oz}^2}\cos\delta. \tag{11.12}$$

To derive this expression for the orientation of the ellipse we start with the general equation (11.10) describing the E-vector for elliptically polarized light,

$$\frac{E_z^2}{E_{oz}^2} + \frac{E_y^2}{E_{oy}^2} - \frac{2E_zE_y}{E_{oz}E_{oy}}\cos\delta - \sin^2\delta = 0,$$

where δ is the phase shift of E_{oz} relative to E_{oy}.

Writing $E_{oz} = a$ and $E_{oy} = b$,

$$\frac{E_z^2}{a^2} + \frac{E_y^2}{b^2} - \frac{2E_zE_y}{ab}\cos\delta - \sin^2\delta = 0. \tag{11.13}$$

This is a general equation for an ellipse shown schematically in figure 11.6.

The ellipse is at an angle, θ, to the z-axis and, when the locus of the point R on the ellipse is at the extremum of the major axis,

$$\tan\theta = \left|\frac{E_z}{E_y}\right|_{\text{maximum}}.$$

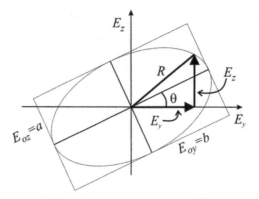

Figure 11.6. Ellipse with minor axis, a, and major axis b, oriented at angle, θ, to the y-axis. The instantaneous resultant of the y- and z-components of the field is R and the tip of R sweeps out the ellipse shown.

Therefore, we find the major axis by finding the maximum value of $E_y^2 + E_z^2 = R^2$. Differentiating this equation and setting the result equal to zero for maximum,

$$2E_y dE_y + 2E_z dE_z = 0. \tag{11.14}$$

Differentiating equation (11.13),

$$2\frac{E_z}{a^2}dE_z + 2\frac{E_y}{b^2}dE_y - \left(\frac{2E_y dE_z + 2E_z dE_y}{ab}\right)\cos\delta = 0$$

$$\left(\frac{E_z}{a^2} - \frac{E_y \cos\delta}{ab}\right)dE_z + \left(\frac{E_y}{b^2} - \frac{E_z \cos\delta}{ab}\right)dE_y = 0. \tag{11.15}$$

Comparing coefficients in equations (11.14) and (11.15),

$$E_y = \frac{E_y}{b^2} - \frac{E_z \cos\delta}{ab} \quad \text{and} \quad E_z = \frac{E_z}{a^2} - \frac{E_y \cos\delta}{ab}.$$

Rearranging these equations leads to,

$$1 = \frac{1}{b^2} - \frac{\cos\delta}{ab}\frac{E_z}{E_y} \quad \text{and} \quad 1 = \frac{1}{a^2} - \frac{\cos\delta}{ab}\frac{E_y}{E_z}.$$

Combining these last two equations,

$$\frac{1}{a^2} - \frac{1}{b^2} = \left(\frac{E_y}{E_z} - \frac{E_z}{E_y}\right)\frac{\cos\delta}{ab}.$$

Now using $\tan\theta = \frac{E_z}{E_y}$,

$$\frac{b^2 - a^2}{a^2 b^2} = (\tan\theta - \cot\theta)\frac{\cos\delta}{ab}.$$

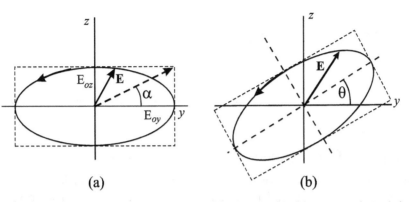

Figure 11.7. Elliptically polarized light (a) axes aligned with y-, z-axes, (b) with axes at angle θ, relative to the y-,z-axes.

and using the trigonometric identity for $(\cot 2\theta)$,

$$\frac{b^2 - a^2}{ab} = (2 \cot 2\theta)\cos \delta,$$

which yields,

$$\tan 2\theta = \frac{2ab \cos \delta}{b^2 - a^2}.$$

Hence we obtain equation (11.12),

$$\tan 2\theta = \frac{2E_{oy}E_{oz}}{E_{oy}^2 - E_{oz}^2} \cos \delta. \tag{11.12}$$

Thus, we see that in order to specify elliptically polarized light we need to identify the ratio of the major and minor axes of the ellipse defining the degree of ellipticity. When this ratio is unity we have circularly polarized light, i.e. a special case of elliptical polarization. We also need to specify the orientation of the ellipse in space, i.e. the direction of the major/minor axes relative to some observer's coordinate frame, by the angle, θ, as shown in figure 11.7 and given by equation (11.12).

11.2 Transformation and analysis of states of polarization

As δ varies from $0 \rightarrow 2\pi$ the polarization varies from linear to elliptical and back to linear. Thus, we may transform the state of polarization between linear and elliptical or vice versa by altering the relative phase of the two components. This is shown schematically in figure 11.8. Changing the relative phase can be done using a material that has a different phase velocity or refractive index for the two components of the E-vector, i.e. the medium will have a different refractive index for two orthogonal directions of polarization, a *birefringent* material. Before discussing the practicalities of transforming or analysing the polarization state we therefore turn to a brief description of the optics of birefringent materials.

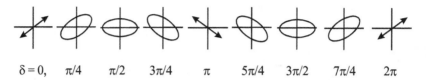

$$\delta = 0, \quad \pi/4 \quad \pi/2 \quad 3\pi/4 \quad \pi \quad 5\pi/4 \quad 3\pi/2 \quad 7\pi/4 \quad 2\pi$$

Figure 11.8. General elliptical state of polarization for different values of relative phase, δ, between the components.

11.3 Optics of anisotropic media; birefringence

Firstly, some background information about electric fields in dielectric media. The optical properties of a material are determined by how the electric displacement field, **D**, inside the medium is related to an 'initial' electric field, **E**, incident 'from outside'. This initial field could originate from a free charge, or a collection of charges, and is the field experienced by a test charge in a vacuum. If the test charge is inside a dielectric material the field, **D**, that it experiences is modified by the surrounding material and how this material is affected by the initial field. The effect of the external field is to polarize the molecules in the medium by inducing a displacement of the electrons relative to the positive nuclear charges. This polarization arising from the induced dipoles produces an additional field—the polarization field, **P**,

$$\mathbf{P} = \varepsilon_0 \chi \mathbf{E} \tag{11.16}$$

where ε_0 is the permittivity of free space (the vacuum) and χ is the susceptibility tensor of the medium. The susceptibility is, in general, a tensor because the ease with which the applied field displaces the molecular charges in one direction depends on the restoring forces in the material in all three orthogonal directions and these may be different. In *isotropic* media these forces are equal in all directions and so the susceptibility can be represented by a scalar quantity. Many crystalline dielectric materials are, however, *anisotropic* and so their optical properties will be different for light polarized in different directions with respect to the axes of the crystal.

This electrical polarization field is proportional to **E** and acts together with the original field to produce the resultant displacement field in the medium, **D**,

$$\mathbf{D} = \varepsilon_0 \mathbf{E} + \mathbf{P} = \varepsilon_0 \varepsilon_r \mathbf{E} \tag{11.17}$$

where ε_r, the relative permittivity, is a tensor, $\varepsilon_r = (1 + \chi)$, and so the components of **D** are related to the components of **E**, in general, by,

$$\begin{pmatrix} \mathbf{D}_x \\ \mathbf{D}_y \\ \mathbf{D}_z \end{pmatrix} = \varepsilon_0 \begin{pmatrix} \varepsilon_{xx} & \varepsilon_{xy} & \varepsilon_{xz} \\ \varepsilon_{yx} & \varepsilon_{yy} & \varepsilon_{yz} \\ \varepsilon_{zx} & \varepsilon_{zy} & \varepsilon_{zz} \end{pmatrix} \begin{pmatrix} \mathbf{E}_x \\ \mathbf{E}_y \\ \mathbf{E}_z \end{pmatrix}. \tag{11.18}$$

It is easy to see now that the x-component of **D** will, as a result of the off-diagonal elements, depend on the strength of the field in the y- and z-directions as well as that in the x-direction; $\mathbf{D}_x = \varepsilon_0(\varepsilon_{xx}\mathbf{E}_x + \varepsilon_{xy}\mathbf{E}_y + \varepsilon_{xz}\mathbf{E}_z)$, and similarly for the other components.

The permittivity tensor matrix can be represented by an ellipsoidal surface such that, with respect to an arbitrary frame of reference defined by x, y, z-axes, the permittivity may be found by the distance from the origin to the surface in the direction of the field. Such a surface is shown in figure 11.9 in which the axes of symmetry of the ellipsoid, X, Y, Z are rotated at an arbitrary angle to the reference frame x, y, z. Since the reference frame is arbitrary, we might just as well choose it such that it matches the axes of symmetry of the material; in which case we can write equation (11.18) as,

$$\begin{pmatrix} \mathbf{D}_x \\ \mathbf{D}_y \\ \mathbf{D}_z \end{pmatrix} = \varepsilon_o \begin{pmatrix} \varepsilon_X & 0 & 0 \\ 0 & \varepsilon_Y & 0 \\ 0 & 0 & \varepsilon_Z \end{pmatrix} \begin{pmatrix} \mathbf{E}_x \\ \mathbf{E}_y \\ \mathbf{E}_z \end{pmatrix}. \tag{11.19}$$

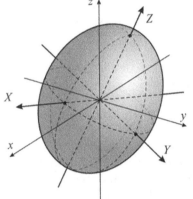

Figure 11.9. An ellipsoidal surface with axes of symmetry X, Y, Z at some angle to the reference frame with axes x, y, z.

This nicely illustrates the mathematical 'trick' of diagonalizing a matrix by a unitary transformation equivalent to rotating the ellipsoid by two successive rotations, e.g. rotate to align the Z-axis with z thus bringing the X and Y axes into the xy-plane, then a rotation about the z-axis to align the X, Y axes with x, y, respectively. We can use this simplified form for the permittivity tensor and remember also that the permittivity, ε_r, of the medium is also equal to the square of the refractive index, n^2. We can therefore construct a surface for which the distance from the origin to the surface represents the refractive index for electric fields, $\mathbf{E}$, in this direction—this is the *refractive index surface*, sometimes known as the *index ellipsoid*.

An isotropic medium is represented by,

$$\varepsilon_X = \varepsilon_Y = \varepsilon_Z \quad (= n^2)$$

Thus, in the isotropic case, the index ellipsoid will be a spherical surface. Anisotropic media will have an index surface that is ellipsoidal. We will consider a particular type of anisotropic medium—usually crystalline—represented by,

$$\varepsilon_X = \varepsilon_Y \neq \varepsilon_Z.$$

There are, in this type of crystal, different values of refractive index for light with its E vectors along different axes,

$$n_X^2 = n_Y^2 \neq n_Z^2.$$

These materials are characterised basically by two values of refractive index and this property is known as *birefringence*.

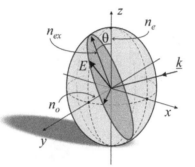

Figure 11.10. A wave propagating with wave vector k, such that the E-vector lies in a plane intercepting the index ellipsoid giving the section shown as the darker shaded ellipse with axes n_o and n_{ex}. The E-vector has one component in the x, y-plane experiencing refractive index, n_o, and an orthogonal component experiencing index, n_{ex}. The value of n_{ex} will vary between n_o and n_e depending on the angle, θ, to the z-axis.

We can now consider how these different values of the refractive index affect the optics of birefringent media. Remembering that we have now 'aligned' our reference axes with the crystal axes, the x, y and z-axes will lie along the axes of crystal symmetry. If a ray of light is polarized such that the E-vector lies in the xy-plane it experiences a refractive index, $n_x = n_y = n_o$, the ordinary index. On the other hand, if the E-vector lies parallel to the z-axis then it experiences an index, $n_z = n_e$, the extra-ordinary index. In general, a wave may propagate through the crystal at any angle to the axes, as shown in figure 11.10, such that the E-vector lies in a plane that intercepts the index ellipsoid in an ellipse. The two components of the E-vector lie on the major and minor axes of this ellipse and experience different refractive indices. The wave consequently splits into two waves travelling with different phase velocities. The component that lies in the x, y-plane is known as the *ordinary ray*, or *o*-ray, and experiences a refractive index n_o—the ordinary index. The orthogonal component, the *extra-ordinary ray*, or *o*-ray, 'sees' a refractive index, n_{ex}, that will vary depending on the direction of the wave vector defined by the angle, θ, to the z-axis.

We can now consider two special cases as shown in figure 11.11. Firstly, we have a ray propagating along the z-axis shown in figure 11.11(a), where the E-vector lies entirely in the x, y-plane and so is an o-ray. Alternatively, as shown in figure 11.11(b), the ray may travel in the x, y-plane and so has its E-vector along the z-axis resulting its being an e-ray.

In the first case, propagation along the z-axis, the direction of the E-vector, i.e. its polarization direction, makes no difference to the refractive index. Thus, the z-axis is

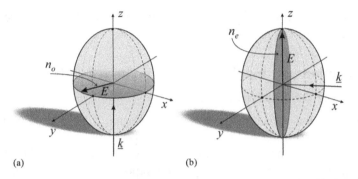

Figure 11.11. Refractive index surface for (a) o-ray propagating along the z-axis with the $\mathbf{E}$-vector in the xy-plane, (b) e-ray propagating in the xy-plane with the $\mathbf{E}$-vector parallel to the z-axis.

an axis of symmetry and is called the *optic axis*. In this case, there is only one axis of symmetry and the crystal is *uniaxial*.

A uniaxial crystal is denoted positive or negative depending on the relative magnitude of the ordinary and extra-ordinary refractive indices; $n_e > n_o$, positive uniaxial and $n_e < n_o$, negative uniaxial. Sections through the index ellipsoids for the two types are shown in figure 11.12.

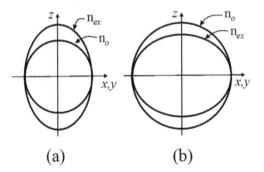

Figure 11.12. Sections through the refractive index surfaces for (a) positive and (b) negative uniaxial birefringent crystals. Note that the sections for o-rays are always circular whereas the sections for e-rays are elliptical.

The difference in refractive indices characterises the degree of birefringence,

$$\Delta n = |n_o - n_e|. \tag{11.20}$$

Since the phase velocity of an o-ray will be the same in all directions the wave front of an o-ray is spherical, whereas the wave front of an e-ray is ellipsoidal as a result of its having different phase velocities in different directions. In general, an unpolarized ray passing through a birefringent crystal in any direction will split into two separate rays—the o-ray and the e-ray. The e-ray, with its elliptical wave front will travel at a slightly different direction to the o-ray owing to the energy flow (i.e. the Poynting vector) not being normal to the wave front in this case. After passing through a parallel-sided crystal the two rays will emerge parallel to each other but slightly displaced and this accounts for the double image seen when an object is viewed through a birefringent material.

11.4 Production and manipulation of polarized light

11.4.1 Modifying the polarization of a wave

Having identified the optical properties of birefringent materials, we now return to the subject of transforming and analysing the polarization states of light. In section 11.2, it was noted that the polarization state of a wave may be modified by changing the relative phase, δ, of the two orthogonal components of the E-field. This can be done by arranging for the two components to travel the same physical distance but at different phase velocities, i.e. they experience different refractive indices so that their relative phase changes with distance travelled. We can do this using a crystal cut with parallel faces normal to the x-axis, i.e. such that the input and output faces of the crystal are parallel to the y, z-plane. A linearly polarized wave travelling in the x-direction, in general, will have components E_y, E_z along the y and z-axes, which experience refractive indices n_o and n_e, respectively, as shown in figure 11.13.

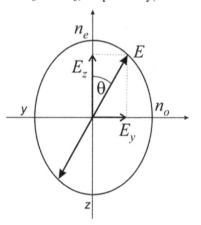

Figure 11.13. Section through the index ellipsoid of a uniaxial crystal for light propagating along the x-axis with the E-vector polarized at angle, θ, to the z-axis. The components E_y and E_z experience refractive indices n_o and n_e, respectively, and so acquire a relative phase shift after propagating through the crystal.

After traversing a length, ℓ, of the crystal a relative phase shift between the two components will be introduced,

$$\delta = \frac{2\pi}{\lambda}|n_o - n_e|\ell. \tag{11.21}$$

For a given piece of birefringent material the value of δ will be determined by the length, ℓ. In the case where the input polarization is linear, E_y and E_z are in phase. In general, as a result of the change in the relative phase, δ, of these components, the output polarization will be elliptical.

The form of elliptical polarization created from a *linearly* polarized input depends on the value of δ and the angle, θ, of the input polarization direction relative to the optic axis (z-axis). These transformations are summarised in table 11.1. For example, the first row of the table describes a linearly polarized input wave with the E-vector at 45° to the y, z-axes and the two components will be given by equation (11.1) with $\delta = 0$. A phase shift of $\delta = \pm\pi/2$, corresponding to a quarter of a wavelength results in the output of circularly polarized light. A crystal producing this phase shift is a *quarter-wave plate*, or $\lambda/4$-plate.

Table 11.1. Transformation of linear polarization by passage through a birefringent plate of different thickness corresponding to quarter- or half-wave retardation in relative phase.

Angle of linearly polarized input	Phase shift introduced by birefringent plate	Output polarization
$\theta = 45°\ (E_y = E_z)$	$\delta = \pm\pi/2$ (Quarter-wave, $\lambda/4$-plate)	Left/right circular
$\theta \neq 45°\ (E_y \neq E_z)$	$\delta = \pm\pi/2$ (Quarter-wave, $\lambda/4$-plate)	Left/right elliptical
$\theta \neq 45°\ (E_y \neq E_z)$	$\delta = \pm\pi$ (Half-wave, $\lambda/2$-plate)	Linear, plane rotated by 2θ

We note that a quarter-wave plate may be used to convert linear to elliptical or vice versa. A plate whose thickness introduces a phase shift of $\delta = \pm\pi$, i.e. a half-wavelength, is known as a half-wave plate, $\lambda/2$-plate. The action of a $\lambda/2$-plate is illustrated in figure 11.14. In this case the optic axis (z-axis) is set at angle, θ, to the direction of the E-vector of the input linearly polarized light. The introduction of a relative phase shift of the y-component results in its pointing in the opposite direction and the rotation of the resultant vector by 2θ.

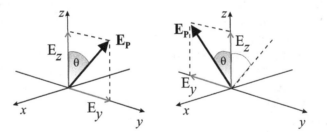

Figure 11.14. Action of a $\lambda/2$-plate with the axis at angle, θ, to the E-vector of plane polarized light shifts phase of one component, E_y, by π relative to the original phase resulting in a rotation by 2θ of the resultant E-vector.

Any given state of elliptically polarized light may be converted to any other desired state of elliptical polarization by passing the light through a sequence of a $\lambda/4$-plate, $\lambda/2$-plate and $\lambda/4$-plate as follows.

The first $\lambda/4$-plate is adjusted so that its optical axis is oriented relative to the original elliptical axes so that the output is linearly polarized.

The axis of the $\lambda/2$-plate is set at an angle, θ, relative to the E-vector to rotate it by 2θ, i.e. θ is chosen to produce linear polarized light at the desired orientation.

Finally, the second $\lambda/4$-plate is rotated relative to the E-vector of the linearly polarized light to achieve the desired elliptical polarization.

11.4.2 Production of polarized light

Polarized light may be produced from unpolarized light using a variety of methods.

    (a) **Fresnel reflection at Brewster's angle**. We saw in section 10.7.1 that p-polarized light has a zero reflection coefficient at Brewster's angle and so the light reflected at this angle of incidence will be linearly s-polarized (see figure 10.9). However, since only about 10% of the incident intensity is reflected this makes a relatively weak source of polarized light. The transmitted light is p-polarized but contains the remainder of the s-polarized component. A series of thin plates at Brewster's angle can be used to reflect more of the s-polarized light until the desired purity of p-polarization is obtained.

    (b) **'Polaroid-type' materials**. Some types of crystals, e.g. tourmaline, absorb light of orthogonal polarizations by different amounts, depending on the direction of polarization relative to the crystal axes. By aligning micro-crystals of such material in a plastic matrix a semi-transparent film can be produced that absorbs light on one polarization preferentially. Sunlight reflected from horizontal planar surfaces such as the sea, is partially horizontally polarized, i.e. s-polarized, as a result of the Fresnel reflection conditions. Polaroid sunglasses work by aligning the axis of the film to absorb preferentially this s-polarized light and so reduce the glare from horizontal surfaces.

    (c) **Birefringent prisms**. Unpolarized light incident on a birefringent material will be resolved into an o-ray and an e-ray that have different refractive indices. As a result, they then have a different angle of refraction and different critical angles, θ_c, as shown in figure 11.15. The difference in refraction angle, in principle, will cause the two beams to diverge and separate into two beams of linear but orthogonal polarization. However, in practice, the angular separation is small and it is preferable to separate the beams more widely. This can be done using a prism of birefringent crystalline material cut so that the beam strikes an angled face at an incidence angle, θ_i, where $\theta_i > \theta_c$ for the o-ray and $\theta_i < \theta_c$ for the e-ray (or vice versa.) This is illustrated in figure 11.15(b). The deviation of the transmitted beam may be compensated by use of a second prism as shown.

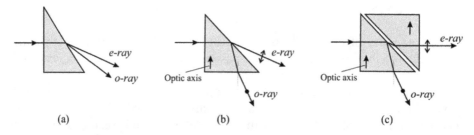

Figure 11.15. Prism polarizers. Unpolarized light is incident normally from the left on the face of a birefringent prism. (a) Both the o-ray and e-ray are transmitted at the output face but at different angles. (b) The angle of incidence on the output face is greater than the critical angle θ_c for the o-ray but less than θ_c for the e-ray. (c) Compensation for the deviation by refraction by use of a second prism.

11.5 Analysis of polarized light

The most general state of light polarization is elliptical. Linear and circular polarizations are special cases of elliptical polarization with $\delta = 0$ and $\delta = \pm \pi/2$ (with $E_{oy} = E_{oz} = E_o$), respectively. We have seen also that linear polarization is a superposition of right- and left-circularly polarized components of equal amplitude.

In order to specify precisely the state of elliptical polarization we need to determine two parameters; the ratio of E_{oy}/E_{oz}, or $\tan \alpha$, and the phase angle, δ. We need to determine the degree of ellipticity and the orientation of the ellipse axes in space relative to the observer's y and z-axes. Figure 11.16(a) shows a general example of elliptically polarized light with axes of symmetry, y', z' tilted at angle, θ, to the observer's y and z-axes.

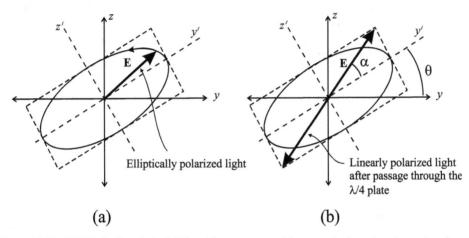

(a)   (b)

Figure 11.16. (a) Elliptically polarized light with axes at an arbitrary angle θ, to the observer's reference y, z-axes. The E-vector is rotating and its tip sweeps out the ellipse circumscribed by the dotted rectangle. (b) Linearly polarized light produced from the elliptically polarized light (a) using a $\lambda/4$-plate aligned with the y', z' axes. The ellipticity is found from $\tan \alpha$.

The E-vector has components along the y', z'-axes with a relative phase, δ, resulting in the rotation of the vector in space. If these components are brought into phase, $\delta = 0$, the light will be linearly polarized with its E-vector at an angle, α, to the y'-axis as shown in figure 11.16(b). This can be achieved using a $\lambda/4$-plate oriented with its optical axis along the major or minor axis of the ellipse. Accordingly, the following method, illustrated in figure 11.17, may be used to specify the state of polarization.

  (i) Pass the light through a linear polarizer. Rotate the linear polarizer to determine approximately the orientation of the major/minor axes of the ellipse—this will be the angle at which a maximum/minimum transmission is obtained. Set the linear polarizer for maximum transmission (figure 11.17(a)).

  (ii) Insert a $\lambda/4$-plate, whose axis is known, **before** the linear polarizer (figure 11.17(b)). Align the optical axis of the $\lambda/4$-plate with the approximate ellipse axis. If it is exactly along the axis then this will result in linearly polarized light. It is then possible to extinguish this light by a suitably oriented linear polarizer.

(iii) Place a linear polarizer in the beam *after* the $\lambda/4$-plate as in figure 11.17(c). Rotate the linear polarizer to check for complete extinction. If total extinction cannot be achieved this means the light was not completely linearly polarized and further adjustment of the $\lambda/4$-plate is required.

(iv) Iterate orientation of the $\lambda/4$-plate and linear polarizer to obtain total extinction. The optical axis of the $\lambda/4$-plate is now at an angle, θ, to the reference y, z-axes. The position of total extinction specifies the orientation of the linearly polarized **E**-vector, which will be at angle, α, to the ellipse axes, y', z'. The angle between this vector and the reference y, z-axes is $(\alpha + \theta)$.

(v) Having now determined θ and $(\alpha + \theta)$, α can be calculated and the phase angle, δ, found from equation (11.12)

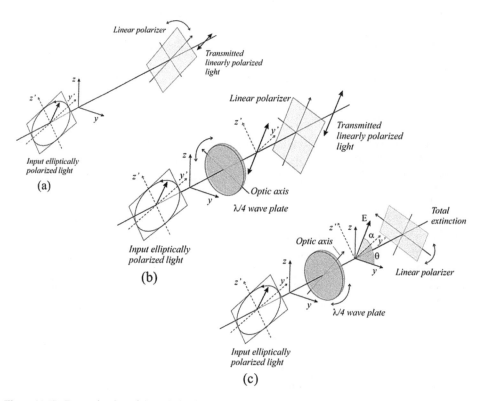

Figure 11.17. Determination of the polarization state for elliptically polarized light using a linear polarizer and a 1/4-plate. See text for details.

11.6 Interference of polarized light

The basic idea of wave interference is that *waves interfere with themselves* not with each other. In order to have a stable interference pattern there must be a fixed phase relationship between the interfering waves, i.e. they must be *coherent*. The light emitted spontaneously, and therefore randomly, from two separate atoms will have no definite phase relationship, i.e. they are *incoherent* and so will not interfere. It is also important to remember that interference occurs between electric fields oscillating in the same direction, i.e. when the wave polarizations are parallel. Therefore, orthogonally polarized waves do not interfere. A dipole source of electromagnetic waves, say an emitting atom, cannot emit, simultaneously, two orthogonally polarized waves. Thus, two orthogonally polarized waves cannot have come from the same source and so cannot interfere. A source, e.g. an atom, may emit a linearly polarized wave that may be resolved into two orthogonal components. These components may interfere if their planes of polarization are made to be the same, e.g. if one component is rotated by a $\lambda/2$-plate to be parallel to the other. Interference is possible because the two components are in phase, i.e. *coherent*.

The plane of polarization of unpolarized light is randomly varying and it might be asked, how can interference be observed using unpolarized light? Interference occurs, for example, in a Michelson interferometer illuminated by light from a thermal, i.e. incoherent, source, because each wave train (photon!) is split into a pair at the beam splitter. Each one of the pair has orthogonal components say, E_{oy} and E_{oz}. The y-component of one of the pair interferes with the y-component of the other one of the pair. Likewise the z-components of the split wave interfere to give the composite interference pattern. Thus, uncorrelated randomly polarized waves from uncorrelated atoms still produce an interference pattern.

The interference properties of polarized light are illustrated in a series of experiments using a Michelson interferometer shown schematically in figure 11.18. The instrument is illuminated by an unpolarized, monochromatic source, e.g. an atomic vapour discharge lamp. In figure 11.18(a) linear polarizers A and B are placed in each of the arms of the interferometer with their transmission axes at right angles. Since the light now reaching the detector from each of the arms is orthogonally polarized, there is no interference.

In figure 11.18(b) the light from the source is linearly polarized by transmission through a linear polarizer C, such that the plane of polarization is at 45° to the vertical and horizontal axes of the polarizers A and B. Consequently, equal intensity and phase-correlated components are transmitted by the polarizers A and B. However, because they are orthogonally polarized, no interference is observed.

In figure 11.18(c) we find something strange. The situation is the same as that in figure 11.18(b), where no interference was observed, except that a linear polarizer D, at 45°, is now placed in the detector arm and interference is observed! The explanation is that the waves (or photons) arriving at polarizer D were phase correlated but orthogonally polarized. Polarizer D selects components of these phase-correlated waves that have parallel polarizations and so they can interfere.

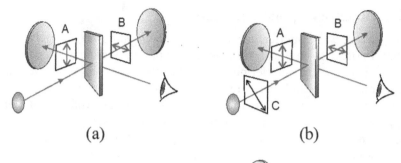

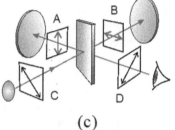

Figure 11.18. A two-beam (Michelson) interferometer illuminated by an unpolarized source. A and B are linear polarizers, i.e. pass only light polarized in the directions shown. (a) Light in paths A and B is orthogonally polarized; no interference. (b) Linear polarizer C at 45° produces phase-correlated components passed by A and B. The components are, however, orthogonally polarized and so no interference is produced. (c) Linear polarizer D at 45° transmits phase-correlated components from polarizer C that are parallel and so interference is produced.

It is worth noting that unpolarized light cannot be fully coherent and so cannot be perfectly monochromatic. Random variation in the plane of polarization results in a random variation of the amplitude along a given axis in space, e.g. the y-axis. This is essentially an amplitude modulated wave and so must contain Fourier components, i.e. other frequencies. Unpolarized light is therefore not purely monochromatic or fully coherent.

These effects are also observed at very low intensities corresponding to a stream of isolated single photons. The wave function for the photons may give probabilities for their being in super-positions of orthogonal states. These and similar effects form the basis of quantum optics... but that is another story.

CPSIA information can be obtained
at www.ICGtesting.com
Printed in the USA
FSHW022159071119
63765FS